INTRODUCTION TO
TOPOLOGY

Krina Shah

TAP

Toronto Academic Press

INTRODUCTION TO TOPOLOGY

Krina Shah

Toronto Academic Press
4164 Lakeshore Road
Burlington ON L7L 1A4
Canada
www.tap-books.com
Email: orders@arclereducation.com

ISBN: 978-1-77956-310-1

ABOUT THE AUTHOR

Krina Shah is a dedicated mathematics tutor with an extensive background in algebra and calculus. She is currently employed in a tutoring role where she rigorously analyzes word problems, interprets data reports, and engages in research activities. Krina holds a Master's degree from the State University of Gujarat and is deeply committed to continuous learning. Her dedication to professional development is reflected in her active participation in various extracurricular activities. In addition to her professional commitments, Krina devotes her time to volunteering with organizations, including blood donation camps, and teaching children with disabilities. Her passion for mathematics, combined with her broad range of experiences, contributes to her development as a compassionate and skilled educational professional.

Table of Contents

List of Figures

PREFACE

Topology is a branch of mathematics that examines the fundamental properties of geometric shapes and spaces, focusing on concepts such as continuity, proximity, and connectivity. It explores the intrinsic properties of objects that remain unchanged under continuous deformations, emphasizing the study of spaces that preserve essential geometric characteristics despite transformations like stretching or bending. By abstracting away specific geometric details while retaining essential structural features, topology provides a powerful framework for understanding spatial relationships and the nature of shapes in a broad and flexible manner. It finds applications across various fields, including physics, computer science, biology, and economics, where understanding the underlying structure and connectivity of spaces is crucial for modeling and analysis.

This book comprises eight chapters that delve into the fundamental concepts and applications of topological spaces. Beginning with Chapter 1, foundational aspects of topological spaces are discussed, exploring the notions of open sets, continuity, and fundamental topological properties that characterize these spaces. Chapter 2 extends this exploration by examining continuity in the context of product topologies, illustrating how topological structures combine to form new spaces essential for understanding product spaces and mappings. Chapter 3 shifts focus to the pivotal concepts of compactness and connectedness, exploring their implications across various topological contexts. In Chapter 4, readers encounter separation axioms, which refine our understanding of how spaces can be distinguished based on the separation of points and sets.

Chapter 5 delves into countability, investigating how topological spaces can be characterized by the cardinality of their bases and covering properties. A profound extension of compact spaces that enriches our understanding of spatial completeness and boundary conditions in topology is discussed in Chapter 6. In Chapter 7, the focus shifts to homotopy and the fundamental group, exploring topological invariants that capture essential properties of spaces through continuous deformations and mappings. These concepts are foundational in algebraic topology, connecting abstract topological spaces to algebraic structures that reveal deeper insights into their topological properties.

Chapter 8 concludes the book by delving into advanced topics in topology, offering a glimpse into specialized areas such as dimension theory, manifold theory, and the interplay between topology and other branches of mathematics. This comprehensive exploration is tailored for students, researchers, and enthusiasts seeking a rigorous yet accessible introduction to the vibrant world of topology, fostering a deeper understanding and appreciation of its theoretical underpinnings and practical applications across diverse scientific disciplines.

—Author

CHAPTER 1

Topological Spaces

LEARNING OBJECTIVES

After studying this chapter, you will be able to:

1. Understand the fundamental concepts of metric spaces and their application in defining distances between points.
2. Explore the properties of continuity and convergence within metric spaces, including sequential and epsilon-delta definitions.
3. Identify and differentiate between various topologies, including the discrete, indiscrete, and order topologies.
4. Examine the construction of topological spaces through bases and subbases, and their role in defining open sets.
5. Analyze the relationships between different topologies using basis comparisons and equivalent metrics.
6. Investigate the concepts of closure, interior, boundary, and derived sets in subspaces of topological spaces.
7. Apply knowledge of neighborhood bases to understand local properties of points in topological spaces.

KEY TERMS FROM THIS CHAPTER

Metric Space

Convergence

Basis

Closure

Neighborhood Base

Continuity

Topology

Subspace

Interior

UNIT INTRODUCTION

Topology has its roots in analysis and geometry, where it extends classical concepts into the framework of modern point-set topology. We begin by revisiting metric spaces, which play a crucial role in establishing foundational concepts such as open sets, closed sets, and neighborhoods. These concepts form the basis for constructing a topology on any set (Kelley et al., 1963).

Our investigation proceeds to derived notions such as the interior, closure, boundary, and limit points of sets within a given space. These ideas help to elucidate the structure and characteristics of subsets in the broader framework of topology (Shabir & Naz, 2011).

Next, we consider the essential concepts of base and subbase, which are crucial for describing a topology on a space (X). A base consists of collections of subsets whose unions define the open sets of the topology, while a subbase generates the topology through intersections of these sets (Fomin, 1943).

Furthermore, we explore the subspace topology that arises when considering subsets of a given topological space. This involves analyzing how the structure of the space influences the topology of its subsets, including how various properties of the overall space manifest within these subsets (Bing, 1951).

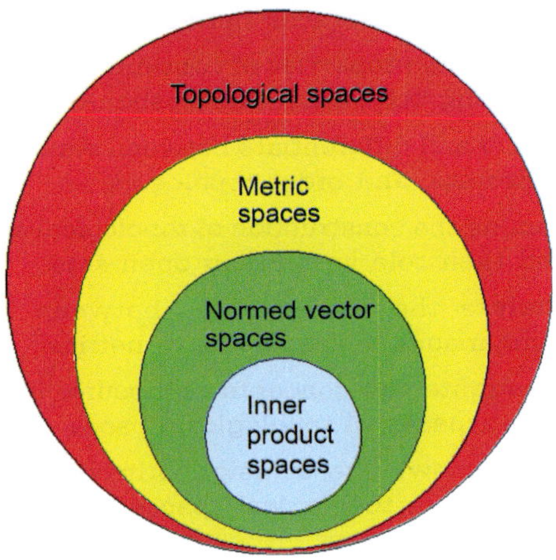

Figure 1.1. Illustration of topological spaces and metric spaces (Source: Math Stack Exchange, Creative Commons License)

1.1. METRIC SPACES

"The fundamental concepts in analysis involve continuity in functions and the convergence of sequences. Both notions depend on an abstract understanding of the proximity of points within a set, which is quantified by the distance between them.

In any set where a concept of distance is well-defined, one can analyze the convergence of sequences and the continuity of functions within these sets. In 1906, Maurice Fréchet introduced the concept of 'metric spaces,' formalizing this idea.

Definition 1.1.1 Consider a nonempty set (X). A metric on (X) is a function $(d: X \times X \to R)$ such that for all $(x, y, z \in X)$, the following conditions hold:

1. Positivity: $d(x, y) \geq 0$ with equality if and only if $(x = y)$,
2. Symmetry: $d(x, y) = d(y, x)$, and
3. Triangle Inequality: $d(x, z) \leq d(x, y) + d(y, z)$.

The set X together with a metric d is called a metric space. The distance among two points, represented as $\left[x, y \in X\right]$ in a set X, is known as d (x, y) (Gruenhage, 1984).

Example 1.1 A significant example of metric space is a Euclidean n-space R^n. The points in this set consist of n-tuples of real numbers, denoted as $x\ (x_1, \ldots, x_n)$. The metric upon this set can be defined by

$$d(x, y) = \sqrt{\sum_{i=1}^{n} (x_i - y_i)^2}$$

To demonstrate that d me satisfies the conditions of Definition 1.1.1, it is necessary to revisit the definition for the "inner product" (or "scalar product") in R^n. This function maps pairs of real numbers $x, y = \sum_{i=1}^{n} x_i y_i$ to the sum of their component-wise products. It operates on vectors in $R^n \times R^n$ into R and returns a real number. The norm for x within the vector space $x \in R^n$ is given by

Remember

Remember the concept of metric spaces and how they provide a framework for understanding continuity and convergence in topology.

$$|x| = \left(\sum x_i^2\right)^{1/2}$$

The subsequent characteristics of $\| x \|$ the absolute value might be easily verified

i. $\|x\| > 0$ for $x \neq 0$;

ii. $\|ax\| = |a| \|x\|$;

iii. $|x, y| \leq |x||y|$ (Cauchy–Schwarz inequality); and

iv. $\|x + y\| \leq \|x\| + \|y\|$

for all x, y ∈ $\mathbb{R}^n$ and a ∈ $\mathbb{R}$.

Verifying d as a metric on R^n, is easy using $d(x, y) = \|x - y\|$. R^1, or Euclidean space, is often denoted by R (Çakallı et al., 2012).

Example 1.1.2 Let C be the field of complex numbers and H be the skew field of quaternions. Let F denote one of these fields. Then the set F^n of ordered n-tuples $(x_1, \ldots, x_n), x_i \in F$, is a vector space under the coordinatewise addition and scalar multiplication. For technical reasons (to be observed later), we will consider F^n as a right vector space over F. Let $\bar{a}$ denote the complex (resp. quaternionic) conjugate of a in $\mathbb{C}$ (resp. H). The standard inner product on F^n is defined by $\langle x, y \rangle = \sum_{i=1}^{n} x_i y_i$for $x = (x_1, \ldots, x_n), y = (y_1, \ldots, y_n)$, and it has the following properties (Bonsangue et al., 1998):

a. $x, y + z = x, y + x, z$;

b. $x, ya = x, y$ a, ±, $= \bar{a}$ (x, y)

c. $x, y = y, x$;

d. x $\neq 0 \Rightarrow x, x > 0$.

We define a function $\|.\| : F^n \to R$ by setting $\| x \| = \sqrt{\langle x, x \rangle}$ and refer to it as the Euclidean norm on Fn. It is easily seen that the function has the properties analogous to those of the Euclidean norm on $\mathbb{R}^n$, and hence there is a metric d on Fn given by $d(x, y) = \|x - y\|$, x, y F^n. We call the metric space the n- dimensional complex (or unitary) space, and the metric space Hn the n-dimensional quaternionic (or symplectic) space.

Example 1.1.3 In the standard Hilbert space 42, the points are infinite real sequences X=(x_i) satisfying $\sum_{i=1}^{n} x_i^2 < \infty$and its metric is defined by

$$d(x, y) = \sqrt{\sum_i (x_i - y_i)^2}.$$

We observe that $d(x,y)$ is always finite. For each positive integer n, we have

$$\sqrt{\sum_{i=1}^{n}(x_i-y_i)^2} \le \sqrt{\sum_{i=1}^{n}x_i^2}+\sqrt{\sum_{i=1}^{n}y_i^2} \le \sqrt{\sum_{i=1}^{n}x_i^2}+\sqrt{\sum_{i=1}^{n}y_i^2} < \infty.$$

The partial sums being bounded, this monotone, nondecreasing sequence must con- verge and one obtains

$$\sqrt{\sum_{i=1}^{n}(x_i-y_i)^2} \le \sqrt{\sum_{i=1}^{n}x_i^2}+\sqrt{\sum_{i=1}^{n}y_i^2}.$$

This implies that $(d(x,y))$ is finite and indeed forms a metric.

Example 1.1.4: The set $C(I)$ of all continuous real-valued functions on $I=[0,1]$, equipped with the metric

$$d(f,g)=\int_0^1 |f(t)-g(t)|\,dt,$$

is an interesting metric space.

Example 1.1.5: Let (X) be a nonempty set *and* $\mathcal{B}(X)$ be the set of all bounded functions $(X\to\mathbb{R})$. The metric $(d^*)on(\mathcal{B}(X))givenby$ (Naidu et al., 2004).

$$d^*(f,g)=\sup\{|f(x)-g(x)|: x\in X\}$$

is known as the supremum metric.

Definition 1.1.2: Let (X,d_X) and (Y,d_Y) be metric spaces. A function $(f: X\to Y)$ is continuous at $(x\in X)$ if, for every $(\epsilon>0)$, there exists $(\delta>0)$ such that $d_X(x,x')<\delta$ implies $d_Y\left(f(x),f(x')\right)<\epsilon$. The function (f) is called continuous if it is continuous at every point $(x\in X)$.

Definition 1.1.3: In a metric space $((X,d))$, the open (r)–ball with center $(x\in X)$ and radius $(r>0)$ is the set $B(x;r)=\{y\in X \mid d(x,y)<r\}$ (Bukatin et al., 2009).

In the real line $\mathbb{R}$, an open (r–ball is simply the open interval $(x-r, x+r)$, and in the plane $(\mathbb{R}^2)$, it is a disc excluding its boundary.

For a function $(f: X \to Y)$ between two metric spaces (X) and $(Y),(f)$ is continuous if and only if for each open ball $B(f(x);\epsilon)$ centered at $f(x)$, there exists an open ball $B(x;\delta)$centered at (x) such that $f(B(x;\delta)) \subseteq B(f(x);\epsilon)$(Mustafa & Sims 2006).

Example 1.1.6: Consider the function $(f:\mathbb{R}^n), \to \mathbb{R}^n),)$ given by $\left(f(x)=\frac{x}{1+||x||}\right)$, where $(\|x\|=\sqrt{x_1^2+\cdots+x_n^2})$.

$$||f(y)-f(x)|| \leq \frac{\|(y-x)+x(\|x\|-\|y\|)+\|x\|(x-y)\|}{(1+||x||)(1+||y||)} \leq \frac{(1+2\|x\|)}{(1+\|x\|)(1+\|y\|)}(\|y-x\|).$$

It can be shown that (f) maps the open ball $B(x;\delta)$ into the open ball $B(f(x);\epsilon)$, thus (f) is continuous.

Definition 1.1.4: A subset (U) of the metric space (X) is called open if, for every point $(x \in U)$, there exists an open r–ball with center (x) contained within (U).

Theorem 1.1.5: Let (X) be a metric space. Then the union of any family of open sets is open, and the intersection of any finite family of open sets is also open.

Proof: The empty set $(\varnothing)$ and the full space (X) are obviously open. If $\{G_{\text{å}}\}$ is a nonempty collection of open subsets of (X), then $(\cup_\alpha G_\alpha)$ is open. To show that the intersection of two open subsets (G_1) *and* (G_2) is open, consider a point $(x \in G_1 \cap G_2)$. Find positive numbers (r_1) and (r_2) such that $(B(x; r\_i) \subseteq G\_i)\, for\, (i = 1,2)$ and $put(r=\min\{r_1,r_2\})$. Then $(B(x;r) \subseteq G_1 \cap G_2)$, proving that $(G_1 \cap G_2)$ is open (Frank, 1971).

1.1.1. Continuity and Convergence in Metric Spaces

Continuity of functions between metric spaces can be fully described in terms of open sets (Figure 1.2).

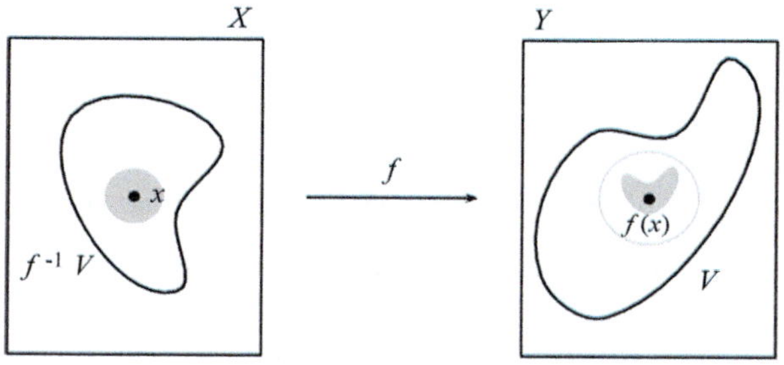

Figure 1.2. Notation in connection with an injective mapping (Source: Joseph Muscat, Creative Commons License)

Theorem 1.1.6: Let (X, d_X) and $((Y, d_Y))$ be metric spaces. A function $(f: X \to Y)$ is continuous if and only if $f^{-1}(G)$ is open in (X) for every open subset (G) of (Y).

Proof:

Suppose $(f: X \to Y)$ is continuous and $(G \trianglelefteq Y)$ is open.

1. Forward Direction (f continuous implies $f^{-1}(G)$ is open):

If $f^{-1}(G) = \varnothing$, it vacuously satisfies being open in (X).

Now, assume $(f^{-1}(G) \neq \varnothing)$. Let $(x \in f^{-1}(G))$.

Since $(x \in f^{-1}(G))$, we have $(f(x) \in G)$.

Because (G) is open, there exists $(\epsilon > 0)$ such that $(B(f(x); \epsilon) \subseteq G)$

Since (f) is continuous at (x), there exists $(\delta > 0)$ such that $f(B(x; \delta)) \subseteq B(f(x); \epsilon)$

This implies $(B(x; \delta) \subseteq f^{-1}(G))$ hence $f^{-1}(G)$ is open in ( )

2. Converse Direction $((f$^{−1}$(G))$ open implies (f) continuous):

Let $(x \in X)$ and $(\epsilon > 0)$ be given.

Consider $(f^{-1}(B(f(x); \epsilon)))$. Since $f^{-1}(B(f(x); \epsilon))$ is open by hypothesis, there *exists* $(\delta > 0)$ such that $B(x; \delta) \subseteq f^{-1}(B(f(x); \epsilon))$.

Therefore, $f(B(x; \delta)) \subseteq B(f(x); \epsilon)$, showing (f) is continuous at (X)

In the study of metric spaces, several key terminologies are used. Consider a metric space (X, d) with nonempty subsets (A) and (B).

1.1.2. Distance and Diameter in Metric Spaces

Distance between Sets: The distance between (A) and (B) is defined as

$$dist(A,B) = inf\{d(a,b) \mid a \in A, b \in B\}.$$

If $(A \cap B = \varnothing)$, then $(\text{dist}(A,B) = 0)$. Disjoint sets can have zero distance between them. If either (A) or (B) is empty, $(\text{dist}(A,B) = \infty)$. For $(x \in X)$ and $(A \subseteq X)$, the distance from (x) to (A) is $\text{dist}(x, A) = \text{dist}(\{x\}, A)$ (Bozkaya & Ozsoyoglu, 1997).

Diameter of a Set: The diameter of $(diam(A))$ of (A), is defined as $diam(A) = sup\{d(a,a') \mid a, a' \in A\}$. It is conventionally considered to be 0 for the empty set. A set is called bounded if its diameter is finite. The metric (d) is termed bounded if $(\text{diam}(X))$ is finite.

Subspaces: If (X) is a metric space and $(Y \subseteq X)$, the distance function restricted to (Y × Y) forms a metric on $(Y).(Y)$, equipped with this metric, is referred to as a subspace of (X). Hence, any subset of a metric space is itself a metric space.

This construction provides various examples of metric spaces Ketterer, 2015).

a. All subsets of Euclidean space (R^n) and the Hilbert space (l^2).

b. Specific examples include:

c. Closed unit (n)−disc $(D^n = \{x \in R^n \mid |x| \leq 1\})$.

d. Unit (n)−cube $(I^n = \{x \in \mathbb{R}^n \mid 0 \leq x_i \leq 1, i = 1, \ldots, n\})$.

e. The unit interval $(I = [0,1])\, in\, (R^1)$.

f. The $(n-1)$−dimensional unit sphere $S^{n-1} = \{x \in \mathbb{R}^n \mid |x| = 1\}$.

Note: $(S^0 = \{-1,1\})$ is a discrete two-point space (Levenshtein, 1992).

1.2. TOPOLOGIES

In metric spaces, the key concepts such as "limits," "continuity," "connectedness," and "compactness" can be explained and many significant theorems for analysis are shown solely using the concept of "open sets." In 1914, Felix Hausdorff abstracted the basic attributes for open sets and introduced a concept allowing discussion of these ideas without relying on metrics (Ma et al., 2009).

1.2.1. Basic Concepts and Definitions

Definition 1.2.1. Any topological structure, or topology, over a set X is a set of subsets with specific qualities.

a. the intersection of two members of $\mathcal{T}$ is in $\mathcal{T}$;

b. the union of any collection of members of $\mathcal{T}$ is in $\mathcal{T}$; and

c. the empty set ∅ and the entire set X are in $\mathcal{T}$.

A set X endowed with a topological structure $\mathcal{T}$ on it is called a topological space. The elements of X are called points, and the members of T are called the open sets. In general, a topological space should be denoted as a pair (X, $\mathcal{T}$). But, it is customary to use the expression "X is a topological space" or, more briefly, "X is a space" to mean (X, $\mathcal{T}$) without mentioning the topology $\mathcal{T}$ for X each time.

Example 1.2.1 Let X be any set. The family $\mathcal{D}$ of all subsets of X is a topology on X, called the discrete topology; the pair (X, $\mathcal{D}$) is called a discrete space. On the other extreme, the family $\mathcal{I} = \{\varnothing, X\}$ is also a topology on X, called the indiscrete or trivial topology; the pair (X, $\mathcal{I}$) is called an indiscrete or a trivial space.

Example 1.2.2 If X={a, b}, then there are two topologies {∅, {a}, X } and {∅, {b}, X} on X aside from the discrete and trivial ones. The set X with one of these topologies is called the Sierpinski space.

Example 1.2.3 By Theorem 1.1.5, the collection of sets declared "open" in a metric space (X, d) is a topology on X ; this is called the topology induced by the metric d or simply the metric topology. In future when a metric space is mentioned, it will be understood that the space is a topological space with the metric topology. In particular, the metric topology generated by the Euclidean metric on any subset of $\mathbb{R}^n$ will be referred to as the usual topology. Unless otherwise stated, a subset of $\mathbb{R}^n$ is assumed to have the usual topology. Similarly, the topologies on $\mathbb{C}^n$ and $\mathbb{H}^n$ induced by the metrics in Example 1.1.2 are referred to as the usual topologies.

Example 1.2.4 Given any set X, the family of all those subsets of X whose complements are finite together with the empty set forms a topology

$\mathcal{T}_f$ on X, called the cofinite (or finite complement) topology. We call (X, $\mathcal{T}_f$) a cofinite space. Similarly, the family of all those subsets of X whose complements are countable together with the empty set is a topology $\mathcal{T}_c$ on X, called the cocountable topology (or countable complement) topology.

We will encounter more serious examples later. It is obvious that one can assign several topological structures to a given set and these can be partially ordered by inclusion relation. If $\mathcal{T}$ and $\mathcal{T}'$ are two topologies on the same set X, we call $\mathcal{T}'$ finer (or larger) than $\mathcal{T}$ if $\mathcal{T} \subseteq \mathcal{T}'$. In this case, we also say that $\mathcal{T}$ is coarser (or smaller) than $\mathcal{T}'$. The terms "stronger" and "weaker" are also used in the literature to describe the above situation. But there is no agreement on their meaning, so we will not use these terms. It may happen that T is neither larger nor smaller than $\mathcal{T}'$; in this case it is said that $\mathcal{T}$ and $\mathcal{T}'$ are not comparable. Clearly, the trivial topology for a set X is the smallest possible topology on X, while the discrete topology is the largest possible topology.

Did you know?

Did you know that topological spaces can vary widely, from the discrete topology where every subset is open to the indiscrete topology where only the empty set and the whole space are open?

Definition 1.2.2 A subset F of a topological space X is closed if its complement X – F is open.

By De Morgan's laws, we see the following properties of closed sets in any topological space.

Proposition 1.2.3 Let X be a space. Then

a. the union of two closed sets is a closed set;
b. the intersection of any family of closed sets is a closed set; and
c. the entire set X and the empty set $\emptyset$ are closed sets.

1.2.2. Examples of Topological Spaces

Example 1.2.5 In $\mathbb{R}$, any closed interval $[a,b]$ is closed according to the above definition, for $\mathbb{R}-[a,b]$ is the union of open sets $(-\infty$, a) and (b, ∞). The set $\mathbb{Z}$ of integers is closed, but the set $\mathbb{Q}$ of rationals is not closed.

The property (a) in Proposition 1.2.3 implies, by iteration, that the union of any finite number of closed sets is closed. However, this does not extend to infinite unions; for example, the union $\bigcup_{n=1}^{\infty}[1/n, 2]$ is not closed in $(\mathbb{R})$.

Example 1.2.6: In a discrete space, every set is both open and closed.

Example 1.2.7: Consider the set $\mathbb{Z}$ of integers with the cofinite topology. In this topology, a finite subset of $\mathbb{Z}$ is closed but not open, $\mathbb{Z}-\{0\}$ is open but not closed, and the set $\mathbb{N}$ of positive integers is neither open nor closed.

Example 1.2.8: In the Euclidean space $(\mathbb{R}^2),(\mathbb{S}^1)$ (the unit circle) and $(\mathbb{D}^2)$ (the closed unit disk) are closed sets, whereas the set

$$\{(x,y) \mid x \geq 0 \text{ and } y > 0\}$$

is not closed (why?).

These examples suggest that a subset can be both closed and open (called clopen), or it may not be either open or closed.

By the duality between the axioms for topology and the properties of closed sets, a topology for a set (X) can also be described by specifying a family $(\mathcal{F})$ of subsets of (X) satisfying the conditions in Proposition 1.2.3. In fact, the family of complements of the members of $\mathcal{F}$ is a topology for (X) such that $\mathcal{F}$ consists of precisely the closed subsets of (X). This shows that the concept of a closed set can be taken as the primitive notion to define a topology (Von Neumann, 1935).

Definition 1.2.4: If (X) is a topological space and $(x \in X)$, then a set $(N \subseteq X)$ is called a neighborhood (written nbd) of (x) in (X) if there is an open set (U) in (X) with $(x \in U \subseteq N)$. Note that a neighborhood is not necessarily an open set, while an open set is a neighborhood of each of its points. In particular, the entire space (X) is a neighborhood of every point. This implies that a neighborhood need not be "small" as one might think. If (N) itself is open, we will call it an "open neighborhood." This is standard practice, although some mathematicians use the term "neighborhood" only in this sense.

Observe the following properties of the family $(\mathcal{N}_x)$ of all neighborhoods of a point (x) in a topological space (X):

a. $(\mathcal{N}_x)$ is nonempty.

b. (x) belongs to each $(N \in \mathcal{N}_x)$.

c. The intersection of two members of $(\mathcal{N}_x)$ is again in $(\mathcal{N}_x)$.

d. If $(N \in \mathcal{N}_x)$ and $(N \subseteq M \subseteq X)$, then $(M \in \mathcal{N}_x)$.

e. If $(N \in \mathcal{N}_x)$, then $\left(N^\circ = \{y \in N \mid N \in \mathcal{N}_y\}\right)$ is also a member of $(\mathcal{N}_x)$.

The last statement follows from the fact that if (U) is an open set in (X) such that $(x \in U \subseteq N)$, then $(U \subseteq N^\circ)$.

The above properties of neighborhoods in a space can be used to work in the opposite direction; starting with them as the axioms for "neighborhoods," we can determine a topology (Robertson & Robertson, 1980).

Proposition 1.2.5: Let (X) be a set and suppose that, for each $(x \in X)$, we are given a family $(\mathcal{N}_x)$ of subsets of (X), satisfying the conditions (a)–(d) above. Then the collection

$$\mathcal{T} = \{U \subseteq X \mid U \in \mathcal{N}_x \text{ for all } x \in U\}$$

is a topology on (X). If (e) is also satisfied, then $(\mathcal{N}_x)$ is precisely the collection of all neighborhoods of (x) relative to $(\mathcal{T})$.

Proof: The verification of the axioms of topology is routine; we prove the last statement only. If (N) is a neighborhood of the point (x), then $(x \in U \subseteq N)$ for some open set (U). Since for each $(y \in U),(N^\circ = \{y \in N \mid N \in \mathcal{N}_y\})$ is a member of $(\mathcal{N}_y)$. Also, it is obvious from the definition of (U) that $(N^\circ \subseteq U)$. So, by (d), $(U \in \mathcal{N}_y)$ and we see that (U) is open. Therefore (N) is a neighborhood of (x), as desired (Spreen, 1998).

Clearly, a subset (U) of a space (X) is open if and only if (U) is a neighborhood of each of its points. Hence, we see that any two topologies for (X), in which each point of (X) has the same family of neighborhoods, must be identical. Accordingly, the topology given in the preceding proposition is unique when the condition (e) is also satisfied. It follows that even the concept of "a neighborhood of a point" may be used as the primitive notion and the conditions (a)–(e) in Proposition 1.2.5 as the axioms for it to define a topology (Wallman, 1938).

1.3. DERIVED CONCEPTS

Definition 1.3.1: Let (X) be a space and $(A \subseteq X)$. The set

$A^{\circ} = \bigcup \{G \mid G \text{ is open in } X \text{ and } G \subseteq A\}$

is the largest open set contained in (A); it is called the interior of (A) in (X). The notation $(\text{int}(A))$ is also used for (A°) (Mader, 2006).

Example 1.3.1: In the real line $(R), (Q^{\circ} = \varnothing = (R - Q)^{\circ}), and ([a,b]^{\circ} = (a,b))$.

Example 1.3.2: In the space $(R^2), (\text{int}(S^1) = \varnothing) and (\text{int}(D^2) = B(0;1))$.

If (X) is a space and $(A \subseteq X)$, then a point of (A°) is called an interior point of (A). Obviously, a point $(x \in X)$. is an interior point of (A) if and only if (A) is a neighborhood of (x). It is also clear that (A) is open if and only if $(A = A^{\circ})$.

Proposition 1.3.2: Let (X) be a space. Then, for $(A, B \subseteq X)$, we have:

(a) $((A^{\circ})^{\circ} = A^{\circ})$,

(b) $(A^{\circ} \subseteq B \Rightarrow A \subseteq B)$,

(c) $((A \cap B)^{\circ} = A^{\circ} \cap B^{\circ})$,

(d) $(A^{\circ} \cup B^{\circ} \subseteq (A \cup B)^{\circ})$.

We leave the simple proofs to the reader. Notice that the reverse inclusion in (d) does not hold good; this is shown by Example 1.3.1.

Definition 1.3.3: Let (X) be a space and $(A \subseteq X)$. The set

$$[\bar{A} = \bigcap \{F \mid F \text{ is closed in } X \text{ and } A \subseteq F\}]$$

is the smallest closed set containing (A). This is called the closure of (A), sometimes denoted by $(\mathrm{cl}(A))$. A point $(x \in \bar{A})$ is referred to as an adherent point of (A).

Example 1.3.3: In the space $(R),(\overline{(a,b)}=[a,b])$ and $(\bar{Q}=R=\overline{R-Q})$.

Example 1.3.4: In the space $(R^2),(\overline{B(0;1)}=D^2)$.

Example 1.3.5: In a cofinite space $(X),(\bar{A}=X)$ for every infinite set $(A \subseteq X)$

It is readily seen that a subset (A) of a space (x) is closed if and only if $(A=\bar{A})$. We also leave the straightforward proofs of the following proposition to the reader.

Proposition 1.3.4: Let (x) be a space and $(A,B \subseteq X)$ Then:

(a) $(\bar{A}=\bar{\bar{A}})$,

(b) $(A \subseteq B \Rightarrow \bar{A} \subseteq \bar{B})$

(c) (¯{A ∪ B} = ¯{A} ∪ ¯{B}),

(d) $(\overline{A \cap B} \subseteq \bar{A} \cap \bar{B})$

Note that the equality in (d) may fail, as is seen by taking $(A=(-1,0))$ and $(B=(0,1))$ in the real line (R).

Theorem 1.3.5: Let (A) be a subset of a space (X). Then $(x \in \bar{A} \Leftrightarrow U \cap A \neq \emptyset)$ for every open neighborhood $(U) of (x)$.

Proof: If there exists an open set (U) such that $(x \in U)$ and $(U \cap A=\varnothing)$, then $(F=X-U)$ is a closed set which contains A but not (x). Thus $(x \notin \bar{A})$. Conversely, if $(x \notin \bar{A})$, then $(U=X-\bar{A})$ is an open neighborhood of (x) disjoint from (A).(\square)

Definition 1.3.6: Let (A) be a subset of a space (X). A point $(x \in X)$ is a limit point (or accumulation point or cluster point) of (A) if every open neighborhood of (x) contains at least one point of $(A-\{x\})$. The set $(\;')$ of all limit points of (A) is called the derived set of (A).

Example 1.3.6: In the space $(X=\{a,b\})$ with the topology $(\{\varnothing,\{a\},X\}),(\{a\}'=\{b\})$ and $(\{b\}'=\varnothing)$.

Example 1.3.7: In (R), every point of $((0,1))$ is a limit point of $((0,1))$, whereas the set (Z) of integers has no limit points.

Example 1.3.8: Every point of (R^3) is a limit point of the subset (A) of those points all of whose coordinates are rational and, at the other extreme, the subset (B) of points which have integer coordinates does not have any limit points.

Theorem 1.3.7: Let (A) be a subset of a space (X). Then $\left(\bar{A}=A\cup A'\right)$.

Proof: If (x) is neither a point nor a limit point of (A), then there is an open neighborhood (U) of (x) such that $(U\cap A=\varnothing)$. Since (U) is a neighborhood of each of its points, none of these is in $\left(A'\right)$. So (U) is contained in the complement of $\left(A\cup A'\right)$, and it follows that $\left(A\cup A'\right)$ is closed. Hence, $\left(\bar{A}\subseteq A\cup A'\right)$. On the other hand, $\left(A'\subseteq\bar{A}\right)$, by Theorem 1.3.5. As $\left(A\subseteq\bar{A}\right)$ always, we find that $\left(A\cup A'\subseteq\bar{A}\right)$, completing the proof.

Corollary 1.3.8: A set is closed if and only if it contains all its limit points.

Proof: (A) is closed $(\Leftrightarrow A=\bar{A}=A\cup A'\Leftrightarrow A'\subseteq A)$

Definition 1.3.9: Let (A) be a subset of (X). The boundary (or frontier) of (A) is defined to be the set

$$\partial A=\bar{A}\cap\overline{X-A}$$

The notation $(\mathrm{bd}(A))$ is also used for (∂A). A *point* $(x\in\partial A)$ is called a boundary point of (A).

Obviously, (∂A) is identical with $(\partial(X-A))$. Also, it is clear that a point $(x\in X)$ is a boundary point of (A) if and only if each (open) neighborhood of (x) intersects both (A) and $(X-A)$.

Example 1.3.9: In $(R), (\partial [0,1] = \{0,1\})$ and $(\partial Q = R)$.

Example 1.3.10: $In(R^2), (\partial D^2 = S^1)$ and $(\partial S^1 = S^1)$.

Example 1.3.11: Let (A) be the set of all points of (R^3) which have rational coordinates. Then $(\partial A = R^3)$.

Theorem 1.3.10: Let (A) be a subset of a space (X). Then

$\bar{A} = A \cup \partial A.$

Proof: By definition, $(\bar{A})$ contains both (A) and (∂A), and hence their union. Conversely, if $(x \in \bar{A} - A), then (x \in \bar{A} \cap (X - A) \subseteq \partial A)$ and the reverse inclusion follows.

As an immediate consequence of this theorem, we have:

Corollary 1.3.11: A set is closed if and only if it contains its boundary.

Definition 1.3.12: Let (A) be a subset of a space (X). A point $(a \in A)$ is called isolated if $(a \notin A')$. The set (A) is called perfect if it is closed and has no isolated points.

Example 1.3.12: Consider the closed unit interval (I) with the subspace topology. The Cantor set is defined by

$$C = I - \bigcup_{n=1}^{\infty} J_n$$

where (J_n) are the removed open intervals in the construction of the Cantor set. This is the set of all points in (I) whose at least one triadic expansion (base 3) contains no 1's. We observe that (C) is perfect. Clearly, (C) is closed. To see that every point of (C) is a limit point, let $(x \in C)$ be arbitrary and (U) be an open interval containing (x). Choose a sufficiently large integer (n) so that (U) contains a closed interval $([x - 1/3^n, x + 1/3^n])$. Now, find an integer $(k \geq 0)$ such that (x) belongs to a closed interval of the form $([k/3^n, (k+1)/3^n])$. Obviously, one endpoint of this interval is different from (x). Thus

(U) contains a point of (C) other than (x), and (x) is a limit point of (C). We close this section with another terminology for subsets in a space, which will also appear several times in this text.

Definition 1.3.13: A subset (A) of a topological space (X) is called dense (or everywhere dense) if $(\bar{A} = X)$.

Example 1.3.13: In the real line (R), both the set of rational numbers (Q) and the set of irrational numbers $(R-Q)$ are dense.

Example 1.3.14: If (X) is an infinite set with the cofinite topology, then the dense subsets of (X) are its infinite subsets (Banon, 2000).

1.4. BASES

The specification of a topology by describing all of the open sets is usually a difficult task. This can often be done more simply by using the notion of a "generating family" for the topology. In this section, we will study two such concepts (Alcantud, 2020).

Given a set (X) and a family $(\mathcal{S})$ of subsets of (X), it is easily seen that the intersection of the collection of all topologies on (X) which contains $(\mathcal{S})$ (certainly nonempty, for the discrete topology on (X) is one such topology) is a topology, denoted by $(\mathcal{T}(\mathcal{S}))$. Clearly, $(\mathcal{T}(\mathcal{S}))$ is the coarsest topology on (X) containing $(\mathcal{S})$ (Bhattacharya et al., 2019).

Definition 1.4.1: Let (X) be a space with the topology $(\mathcal{T})$. A subbasis (or subbase) for $(\mathcal{T})$ is a family $(\mathcal{S})$ of subsets of (X) such that $(\mathcal{T} = \mathcal{T}(\mathcal{S}))$.

If $(\mathcal{S})$ is a subbasis for the topology $(\mathcal{T})$ on (X), the members of $(\mathcal{S})$ are open in (X) and referred to as subbasic open sets. Furthermore, the topology $(\mathcal{T})$ consists of $(\varnothing),(X)$, all finite intersections of members of $(\mathcal{S})$, and all unions of these finite intersections. This can be easily ascertained by verifying that the collection of these sets is a topology for (X), which contains $(\mathcal{S})$ and is coarser than $(\mathcal{T})$. It follows that the topology $(\mathcal{T})$ is completely determined by the family $(\mathcal{S})$ (Rodabaugh, 1999).

As we have seen above, any family $(\mathcal{S})$ of subsets of a given set (X) serves as a subbasis for some topology on (X), viz., $(\mathcal{T}(\mathcal{S}))$. So, to define a topology on (X), it suffices to specify a family $(\mathcal{S})$ of subsets of (X) as a subbasis. The resulting topology is said to be generated by the subbasis $(\mathcal{S})$. We illustrate this by introducing a topology on an ordered set $(X,\prec)$. For each pair of elements $(a,b\in X)$, define (Fuster & Silvi, 2000).

$(a,b)=\{x\in X \mid a\prec x\prec b\}$ (open interval),

$[a,b]=\{x\in X \mid a\preceq x\preceq b\}$ (closed interval),

$[a,b)=\{x\in X \mid a\preceq x\prec b\}$ (half-open or half-closed interval),

$(a,b] = \{x \in X \mid a \prec x \preceq b\}$ (half-open or half-closed interval).

And, for each $(a \in X)$, we define

$(-\infty, a) = \{x \in X \mid x \prec a\}$ (open rays or one-sided open intervals),

$(-\infty, a] = \{x \in X \mid x \preceq a\}$ (closed rays or one-sided closed intervals).

It is obvious that $(-\infty, a) = (a_0, a)$ if (a_0) is the first (or smallest) element, and $(a, +\infty) = (a, b_0]$ if (b_0) is the last (or largest) element of (X).

We ought to find a topology on (X) which justifies the use of adjectives "closed" and "open" here. Notice that an open interval (a,b) can be obtained as an intersection of the rays $((a,+\infty))$ and $(-\infty, b)$ so that a topology which contains these rays certainly contains (a,b).

1.4.1. Subbases and Bases in Topology

If $(\mathcal{S})$ is a subbasis for the topology $(\mathcal{T})$ on (X), the members of $(\mathcal{S})$ are open in (X) and referred to as subbasic open sets. Furthermore, the topology $(\mathcal{T})$ consists of $(\varnothing)$, (X), all finite intersections of members of $(\mathcal{S})$, and all unions of these finite intersections. This can be easily ascertained by verifying that the collection of these sets is a topology for (X), which contains $(\mathcal{S})$ and is coarser than $(\mathcal{T})$. It follows that the topology $(\mathcal{T})$ is completely determined by the family $(\mathcal{S})$.

As we have seen above, any family $(\mathcal{S})$ of subsets of a given set (X) serves as a subbasis for some topology on (X), *viz.*, $(\mathcal{T}(\mathcal{S}))$. So, to define a topology on (X), it suffices to specify a family $(\mathcal{S})$ of subsets of (X) as a subbasis. The resulting topology is said to be generated by the subbasis (Sharma, 1971).

Definition 1.4.2: Let (X) be an ordered set. The order topology (or the interval topology) on (X) is the topology generated by the subbasis consisting of all the "open rays" $((-\infty, a))$ and $((a,+\infty)), where (a) ranges over (X)$.

In the order topology on (X), an open interval $((a,b))$ is obviously open, and a closed interval $([a,b])$, being the complement of $((-\infty,a)\cup(b,+\infty))$, is closed.

Example 1.4.1: Consider the set (R) of real numbers with the usual order relation on it. Since the open rays $((-\infty,a))$ and $((a,+\infty))$ (where $(a \in \dashv\!\text{o}\{R\})$) are open in the Euclidean topology on (R), the order topology for $(\dashv\!\text{o}\{R\})$ is coarser than the Euclidean topology. On the other hand, every open ball, being an open interval, is open in the order topology. Therefore, every open subset of the real line (R) is open in the order topology, and the two topologies for (R) coincide. Thus, we see that the family of all open rays in $(R)$is a subbase for the topology of the real line (R).

Since the operations of union and intersection both are involved in the construction of a topology from a subbasis, an important simplification occurs if the open sets are constructed only by taking unions of members of $(\mathcal{S})$. This is possible, for example, if $(\mathcal{S})$ is closed under the formation of finite intersections; this situation motivates the following.

Definition 1.4.3: Let $((X,\mathcal{T}))$ be a space. A basis (or base) for $(\mathcal{T})$ is a family $(\mathcal{B}\subseteq\mathcal{T})$ such that every member of $(\mathcal{T})$ is a union of members of $(\ )$

If $(\mathcal{B})$ is a basis for a topology $(\mathcal{T})$ on (X), then $(\mathcal{T})$ is the coarsest topology on (X) containing $(\mathcal{B})$. For, if $(\mathcal{T}')$ is a topology with $(\mathcal{B}\subseteq\mathcal{T}')$, then all unions of members of $(\mathcal{B})$ are certainly in $(\mathcal{T}')$ and so $(\mathcal{T}\subseteq\mathcal{T}')$. We say that the topology $(\mathcal{T})$ is generated by the basis $(\mathcal{B})$. The members of $(\mathcal{B})$ are referred to as the basic open sets in (X), and $(\mathcal{B})$ is also called a basis for the space (X). There is a simple characterization of bases, which some authors use as a definition (Szymański & Turzański, 1982).

Proposition 1.4.4: A collection $(\mathcal{B})$ of open subsets of a space (X) is a basis if and only if for each open subset (U) of (X) and each point $(x \in U)$, there exists a $(B \in \mathcal{B})$ such that $(x \in B \subseteq U)$.

By the preceding proposition, we have a useful way to describe the open subsets of a space (X) when its topology is given by specifying a basis A set $(G \subseteq X)$ is open if and only if for each $(x \in G)$, there is a $(B \in \mathcal{B})$ such that $(x \in B \subseteq G)$. It follows that the topology of a space is completely determined by a basis (Pang & Xiu, 2019).

Example 1.4.2: In a discrete space the family of all singleton sets is a basis.

Example 1.4.3: In a metric space the collection of open balls is a basis for the metric topology on (X).

Observe that a family $(\mathcal{S})$ of subsets of a space (X) is a subbasis if and only if the family of all finite intersections of members of $(\mathcal{S})$ is a basis for (X). This basis is said to be generated by $(\mathcal{S})$. (We remark that some authors elude the convention that (X) is the intersection of the empty subfamily of $(\mathcal{S})$ and require that a subbasis $(\mathcal{S})$ for (X) must also satisfy the condition $(X \in \mathcal{S})$. It follows that the base generated by the subbasis of an ordered space (X) consists of all open rays, all open intervals $((a,b))$, the empty set $(\varnothing)$, and the full space (X). Notice that if $(x \in X)$ is not the first or the last element, then (x) belongs to an open interval in (X), for there exist elements $(a,\ b \in X)$ such that $(x \in (a,+\infty))$ and $(x \in (-\infty,b))$. Moreover, if (X) has no last element, then $((a,+\infty))$ is a union of open intervals, and if (X) has no first element, then $((-\infty,a))$ is a union of open intervals (Li & Li, 2015) (Figure 1.3).

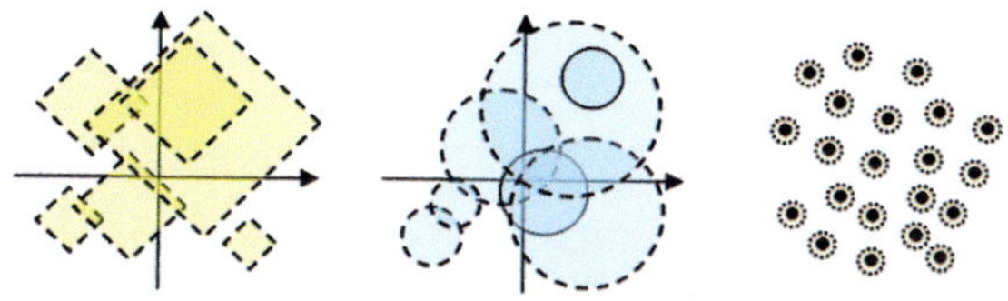

Figure 1.3. Bases and subbases of topology (Source: Math Strek, Creative Commons License)

1.4.2. Example of Order Topology

Example 1.4.5: Recall that if (x) and (y) are two ordinal numbers, then $(x < y)$ means (x) is an element of (y), and (0) denotes the least ordinal number ((Shankar, 1977). Let (Ω)be the first (least) uncountable ordinal number, and consider the set of all ordinal numbers less than (Ω). Obviously, (Ω) is the last element of this set, and it is denoted by $(y < \Omega)$ By definition, the order topology for $(y < \Omega)$ is generated by the subbasis consisting of the right half-open intervals $([0,y))$ for $(y < \Omega)$ and the left half-open intervals $(x,\Omega])$ for $(x < \Omega)$ Accordingly, a basis of the ordered space $([0,\Omega])$ consists of all these intervals and the open intervals (x,y).

Since each ordinal number $(y < \Omega)$ has an immediate successor, this topology for $([0,\Omega])$ is also generated by the basis consisting of $(\{0\})$ and the left half-open intervals (x, y), where $(0 \leq x < y \leq \Omega)$ Notice that if $(x = 0)$ or it has an immediate predecessor in $([0, \Omega])$ then $(\{x\})$ is open (MacKintosh & Lubensky, 1991) (Figure 1.4).

Tip

In the order topology, a basis for the topology is given by intervals (a, b) or [a, b) depending on whether the order is total or partial.

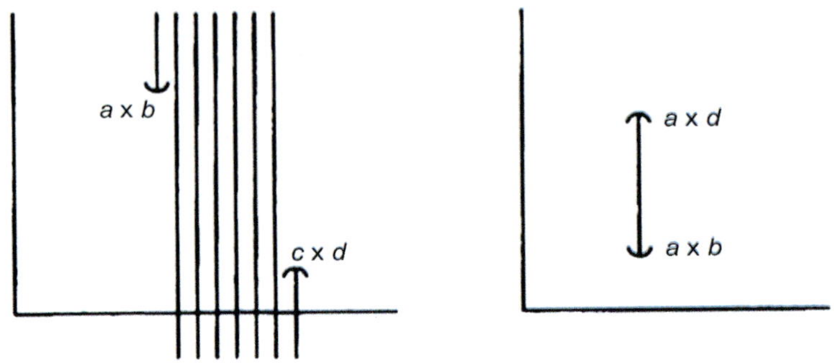

Figure 1.4. Open intervals in the dictionary ordering on (Source: Tej Bahadur, Creative Commons License)

1.4.3. Using Bases to Define Topologies

In the above examples, we have described bases for the given topologies. But, the primary motive behind the concept of basis is to specify a topology on a set (X) by giving a basis. A natural question arises: whether a given family of subsets of (X) would be a base for some topology on (X). The answer to this question is not always positive. For example, the family $(\mathcal{B} = \{\{a,b\},\{a,c\}\})$ of subsets of $(X = \{a, b, c\})$ cannot serve as a basis for a topology on (X), since any topology having $(\mathcal{B})$ as a basis must contain $(\{b\})$, which cannot be expressed as the union of members of $(\mathcal{B})$. So we ought to know when a given collection $(\mathcal{B})$ of subsets of (X) can serve as a basis for some topology on (X). Assume that $(\mathcal{B})$ is a basis for some topology on (X). Then we have $(X = \square\ cB)$, for (X) is open. And, for every pair of sets $(B_1, B_2 \in \mathcal{B})$ and for each

$(x \in B_1 \cap B_2)$, there exists $(B_3 \in \mathcal{B})$ with $(x \in B_3 \subseteq B_1 \cap B_2)$, since $(B\_1 \cap B\_2)$ is open. In fact, these conditions are also sufficient, as we see below (Nicolau et al., 2011).

1.4.4. Theorem on Bases

According to Theorem 1.4.5, let $(\mathcal{B})$ be a collection of subsets within the set (X) that satisfy

1. $(X = \bigcup cB)$,
2. For each pair of sets $(B_1, B_2 \in \mathcal{B})$ for all $(x \in B_1 \cap B_2)$, it represents $(B_3 \in \mathcal{B})$ with $(x \in B_3 \subseteq B_1 \cap B_2)$.

There is a specific way to structure (X) so that (B) serves as a basis.

Proof: Consider $T(B)$ the collection of each set $(U \subseteq X)$ so that for every $(x \in U)$ *thereexists* $(B \in B)$ *with* $(x \in B \subseteq U)$. It is evident that $(T(B))$ includes $(\varnothing)$ (vacuously).

Given $(U_1, U_2 \in T(B)))$, when $(x \in U_1 \cap U_2)$, one may identify $(B_i \in T(B)), (i = 1, 2)$, resulting in $(x \in B_i \subseteq U_i)$. According to hypothesis, $(B_3 \in B)$ exists according to $(x \in B_3 \subseteq B_1 \cap B_2 \subseteq U_1 \cap U_2)$, implying that $(U_1 \cap U_2 \in T(B))$. So, $(T(B))$ represents a topology upon (X). Furthermore, the definition from $(T(B))$ reveals that $(B \subseteq T(B))$ and each member for $(T(B))$ has a union for (B). Thus, (B) is the base for $T(B)$.

Example 1.4.6: Now analyze the family of "right half-open interval $([a,b))$, where $(a, b \in R)$ and $(a < b)$. Verifying that this family satisfies the criteria for becoming a basis is straightforward. The topology created from this basis is referred to as the lower portion over (R).

The set (R_l) alongside this topology is known as the "Sorgenfrey line" and is indicated by $(R\mathbb{I})$. Within this context, all intervals of the type $((a,b))$ *or* $((a,+\infty))$ are considered open, where $((a,b))$ *or* $((a,+\infty))$ and $((a,+\infty) = \bigcup\{[x, x+1) \mid a < x < b\})$ However, it is important to note that these intervals are not classified as closed (McArthur &

Retherford, 1965). In a similar vein, the family $\left((a,b]\right)$, where $\left(a,b\in R\right)$ and $\left(a<b\right)$, gives rise to a topology on $\left(R\right)$ known as the upper limit topology.

By utilizing bases, we can establish a straightforward criterion for comparing two topologies. This criterion is determined by the bases specified for each topology (Ferenczi, 2005).

1.4.5. Comparison of Topologies by Bases

Proposition 1.4.6: Let $\left(T\right)$ and $\left(T_r\right)$ be topologies on a set $\left(X\right)$ generated by the bases $\left(\exists c\{B\}\right)$ and $\left(\mathcal{B}_r\right)$, respectively. Then $\left(T\right)$ is coarser than $\left(T_r\right)$ if and only if for each $\left(B\in\mathcal{B}\right)$ and each $\left(x\in B\right)$, there exists $\left(B_r\in\mathcal{B}_r\right)$ such that $(x\in B_r\subseteq B)$

Proof: ⇒ Given $\left(B\in\mathcal{B}\right)$, we have $\left(B\in T_r\right)$, since $\left(T\subseteq T_r\right)$. If $\left(x\in B\right)$, then there exists $\left(B_r\in\mathcal{B}_r\right)$ with $\left(x\in B_r\subseteq B\right)$, since $\left(\mathcal{B}_r\right)$ is a basis for $\left(T_r\right)$ (Zegura et al., 1997).

(⇐) Let $\left(U\right)$ be $\left(T\right)$–open. If $\left(x\in U\right)$, then there exists $(B\in\mathcal{B})$ with $(x\in B\subseteq U)$, since $(\mathcal{B})$ is a basis for $\left(T\right)$. By our hypothesis, there exists $\left(B_r\in\mathcal{B}_r\right)$ such that $\left(x\in B_r\subseteq B\right)$. So $\left(x\in B_r\subseteq U\right)$. It follows that $\left(U\right)$ is $\left(T_r\right)$–open, and so $\left(T\subseteq T_r\right)$.

1.4.6. Examples Illustrating Proposition

Example 1.4.7: The topology of the real line $\left(R\right)$ is strictly smaller than that of $\left(R_l\right)$ (the lower limit topology) and the upper limit topology. We already know that the open intervals $\left(a,b\right)$ form a basis for the Euclidean topology and the half-open intervals $\left([a,b)\right)$ form a basis for the lower limit topology. Also, for each $\left(x\in(a,b)\right)$, we have $\left(x\in[x,b)\subseteq(a,b)\right)$. So, by the preceding proposition, the Euclidean topology is coarser than the lower limit topology. However, the basis element $\left([a,b)\right)$ for the lower limit topology is not open in the Euclidean topology, since there is no open interval that contains $\left(a\right)$ and is contained in $[a,\ b]$.

Distinct Bases Generating the Same Topology: Notice that distinct bases or subbases may generate the same topology. For instance, the subcollection of open intervals with rational endpoints is also a basis for the usual topology on $\left(R\right)$. Another illustration is given in Figure 1.5.

Example 1.4.8: Consider the two topologies generated by the families of open balls in the metrics $\left(d\right)$ and $\left(d^*\right)$ on $\left(R^n\right)$, given by:

$$d\left((x_i),(y_i)\right)=\sqrt{\sum_{i=1}^{n}\left|x_i-y_i\right|^2}$$

or

$$d^*((x_i),(y_i)) = |x_i - y_i|.$$

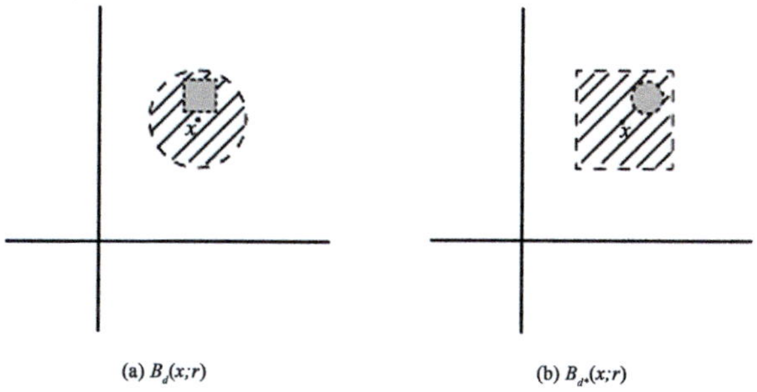

Figure 1.5. Shapes of open balls in the metrics d and d∗ on (Source: Tej Bahadur, Creative Commons License)

Here, (d) represents the Euclidean metric, and (d^*) represents the maximum metric $\left(or\left(L^{\infty}\right)metric\right)$. The open balls in (d) and (d^*) form bases for their respective topologies on (R^n). These two topologies are different, but both are important in their own contexts, with different utility in various mathematical analyses and applications (Rocha & Caires, 2021).

1.4.7. Equivalent Metrics with Bounded Diameter

Corollary 1.4.9: Let $((X,d))$ be a metric space. Then, for each real $(\lambda > 0)$ there is a metric (d_λ) equivalent to ( ) such that the diameter of (X) in (d_λ) is less than λ.

Proof: Define $(d_\lambda(x,y) = \min\{\lambda, d(x,y)\})$ It is easily verified that (d_λ) is a metric on (X) in which $(\mathrm{diam}(X) \le \lambda)$. Clearly, $(B_d(x;\epsilon) \subseteq B_{d_\lambda}(x;\epsilon))$ and, for $(\delta = \min\{\lambda, \epsilon\})$, we have $(B_{d_\lambda}(x;\delta) \subseteq B_d(x;\epsilon))$ Therefore, (d_λ) is equivalent to \(d\), by Proposition 1.4.8 (Kahalon et al., 2022).

1.4.8. Neighborhood Bases

Returning to bases, we see that if $(\mathcal{B})$ is a basis of a space (X), then the neighborhoods of a point $(x \in X)$ are determined by a subfamily of $(\mathcal{B})$, namely, $\left(\dashv c\{B\}\_x = \{B \in \dashv c\{B\} \mid x \in B\}\right)$. For, a set $(N \subseteq X)$ is a neighborhood of (x) in (X) if and only if (N) contains some member of $(\mathcal{B}_x)$. This leads to the following notion (Figure 1.6).

Definition 1.4.10: Let (X) be a topological space and $(x \in X)$. A 'neighborhood basis' (or a local basis) at (x) is a collection $(\mathcal{B}_x)$ of neighborhoods of $(x \backslash)$ in (X) such that each neighborhood of (x) in (X) contains some member of $(\mathcal{B}_x)$.

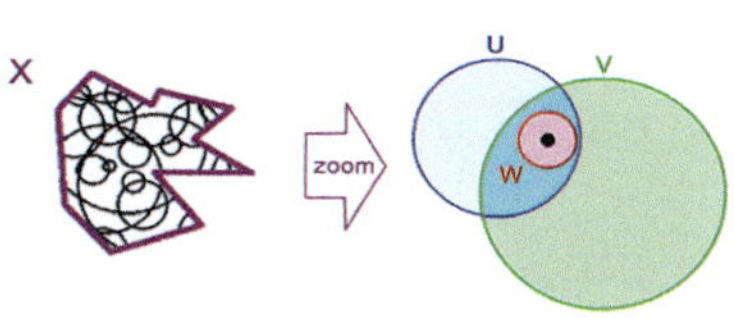

Figure 1.6. Illustration of Bases of neighborhood (Source: Peter Saveliev, Creative Commons License)

Examples:

1. Example 1.4.9: In a topological space, the family of all open neighborhoods of a point is a neighborhood basis at that point.
2. Example 1.4.10: In a discrete space (X), there is a neighborhood basis at a point (x) which consists of just one set, viz., $(\{x\})$.
3. Example 1.4.11: In a metric space (X), the open (r)–balls about a given point (x) for all real $(r>0)$ form a neighborhood basis at (x).

In a space (X), it is easily seen that a collection of neighborhood bases (B_x) at different points (x) satisfies the following conditions:

a. $(\mathcal{B}_x)$ is a nonempty set for each $(x \in X)$;

b. $(x \in B)$ for every $(B \in \mathcal{B}_x)$;

c. for every two sets $(B_1, B_2 \in \mathcal{B}_x)$, there exists a $(B_3 \in \mathcal{B}_x)$ such that $(B_3 \subseteq B_1 \cap B_2)$; and

d. for each $(B \in \mathcal{B}_x)$, there exists a $(B_r \in \mathcal{B}_x)$ such that (B) contains a member of $(\mathcal{B}_y)$ for every $(y \in B_r)$.

Conversely, we have the following theorem.

Theorem 1.4.11: Topology Defined by Neighborhood Bases

Theorem 1.4.11: Let (X) be a set and suppose that, for each $(x \in X), (\mathcal{B}_x)$ is a family of subsets of (X) satisfying the conditions (a)–(d) above. Then there is a unique topology on (X) such that $(\mathcal{B}_x)$ is a neighborhood basis at (x) for each $(x \in X)$.

Proof: Consider the collection

$$\mathcal{T} = \{G \subseteq X \mid G \text{ contains a set } B \in \mathcal{B}_x \text{ for every} x \in G\}.$$

It is easily verified that $(\mathcal{T})$ is a topology for (X). Next, we show that each $(\mathcal{B}_x)$ is a neighborhood basis at (x) in the space $((X, \mathcal{T}))$. Given a set $(B \in \mathcal{B}_x)$, put $(U = \{y \in X \mid B \text{ contains a set } C_y \in \mathcal{B}_y\})$. It is then obvious that $(x \in U \subseteq B)$. Furthermore, (U) is open in the topology $(\mathcal{T})$. For, if $(y \in U)$, then (B) contains a set $(C_y \in \mathcal{B}_y)$, by the definition of (U). And, by (d), there exists a set $(C_y \in \mathcal{B}_y)$ such that (C_y) contains some member of $(\mathcal{B}_z)$ for every $(z \in C_y)$. Since $(C_y \subseteq B)$ we have $(C_y \subseteq U)$ So (U) is open, and thus (B) is a neighborhood of (x). Moreover, if (N) is a neighborhood of (x), then there exists a $(G \in \mathcal{T})$ such that $(x \in G \subseteq N)$. By the definition of $(\mathcal{T}),(G)$ contains a member of $(\mathcal{B}_x)$, and therefore so does (N). Thus we see that $(\mathcal{B}_x)$ is a neighborhood basis at (x).

Finally, to prove the uniqueness of (T), suppose that there is another topology (T_r) on (X) such that for every $(x \in X)$, the family (B_x) is a neighborhood basis at (x). That is, for each $(x \in X)$, the open sets in (T_r) contain a set from (B_x) as a neighborhood.

Now, consider a set $(G \subseteq X)$. We claim that (G) is open in (T_r) if and only if for every point $(x \in G)$, there exists a set $(B \in B_x)$ such that $(B \subseteq G)$. This condition is precisely the definition of the open sets in (T), which means that the topologies (T_r) and (T) must be identical. Hence, $(T_r = T)$, establishing the uniqueness of (T)..The preceding theorem shows that the concept of neighborhood basis can be used as a primitive notion to define a topology on a set (X) (Brown, 1996).

1.5. SUBSPACES

Proposition 1.5.4: If (Y) is a subspace of (X), and (Z) is a subspace of (Y), then (Z) is a subspace of (X).

This property highlights that the subspace construction is transitive. If (Z) is a subset of (Y), and (Y) is a subset of (X), then (Z) inherits a topology from (Y), which in turn inherits its topology from (X) (Halmos, 1969).

1.5.1. Closure and Interior in Subspaces

Proposition 1.5.5: Let (Y) be a subspace of a space (X).

If (Y) is closed (or open) in (X) and (A) is closed (or open) in (Y), then (A) is closed (or open) in (X).

Proof: By Proposition 1.5.2, $(A = Y \cap F)$ for some closed subset (F) of (X). Since the intersection of two closed sets is closed, (A) is closed in (X). A similar argument applies to the "open" case.

These propositions underscore how properties of openness and closedness are preserved or inherited through subspace topologies relative to the parent space (X) (Levine, 1961).

1.5.2. Basis and Neighborhood Base in Subspaces

Proposition 1.5.5: Let (X) be a space and $(Y \subseteq X)$.

1. (a) If (B) is a basis (resp. subbasis) of (X), then $(Y \cap B)$ is a basis (resp. subbasis) for the relative topology of (Y).

2. (b) If (B_x) is a neighborhood base at $(x \in X)$ and $(x \in Y)$, then $(Y \cap B_x)$ is a neighborhood basis at (x) in $(\ )$

This proposition indicates how to construct bases and neighborhood bases for a subspace (Y) of a space (X) using the bases and neighborhood bases of (X) (Zhu et al., 2016).

1.5.3. Closure, Interior, Boundary, and Derived Set in Subspaces

Proposition 1.5.6: Let (Y) be a subspace of a $space(X)$, and $(A \subseteq Y)$.

$$-(\overline{A}^{Y} = \overline{A}^{X} \cap Y)$$

$$-(A_Y^r = A_X^r \cap Y)$$

$$-(A_Y^\circ \supseteq A_X^\circ \cap Y = A_X^\circ)$$

$$-(\partial A_Y \subseteq \partial A_X \cap Y)$$

Proof: $(\overline{A}^X)$ is closed in (X), so $(\overline{A}^X \cap Y)$ is closed $in(Y)$ and contains (A). Therefore, $(\overline{A}^Y \subseteq \overline{A}^X \cap Y)$. Conversely, $(\overline{A}^Y)$ is closed in (Y) and equals $(Y \cap F)$ for some closed (F) $in(X)$, hence $(A \subseteq F)$, and $(\overline{A}^X \cap Y \subseteq F \cap Y = \overline{A}^Y)$. Similar arguments apply to the other sets.

Example 1.5.4: Relative Topology and Order Topology

Consider $(Y = (1,2) \subseteq R)$ with the usual order and topology.

a. ( ) is open in (Y) because it does not belong to (Y).

b. In the order topology of (Y), any basic open set containing (0) extends beyond (1), showing the difference in topologies between (Y) and (R).

Example 1.5.5: Relative Topology vs. Order Topology

Let (J) be an interval in an ordered space (X).

a. The sets $((-\infty,a) \cap J)$ and $\backslash((a,\infty) \cap J)$, where $(a \in X)$, form a subbasis for the relative topology on (J).

b. An open ray in (J) under the order topology is $((-\infty,a) \cap J)$ or $((a,\infty) \cap J)$, indicating that the relative topology is finer than the order topology on (J), and they coincide.

These examples illustrate how the relative topology of a subspace can behave differently from the topology induced by the order structure, depending on the specific subset $(Y \subseteq X)$ (Aleman & Korenblum, 2008).

ACTIVITY 1.1.

Explore the concept of subspaces in topology by considering how closures, interiors, boundaries, and derived sets behave differently in subspaces compared to the whole space.

SUMMARY

Chapter 1 introduces the foundational concepts essential to the study of topology. It begins by defining metric spaces and discussing the notions of continuity and convergence within these spaces. This discussion establishes the basis for understanding how distances between points are quantified and how sequences approach their limits.

The chapter then transitions to the study of topologies, extending beyond the scope of metric spaces. It introduces fundamental concepts and definitions, including open sets, closed sets, and neighborhood bases. These ideas are illustrated through examples that demonstrate the diverse ways in which topologies can be defined and analyzed.

A significant focus of the chapter is the role of bases and subbases in topology. The text explains how these collections are used to construct topologies and how they serve as a framework for understanding open sets and neighborhood systems. Through detailed examples, readers learn how a basis can uniquely determine a topology and how different bases can generate equivalent topologies.

The chapter also examines subspaces of topological spaces, addressing how concepts such as closure, interior, boundary, and derived sets behave within these subsets. The relationship between the topology of a subspace and its parent space is explored, with an emphasis on the continuity of topological properties across different scales.

MULTIPLE CHOICE QUESTIONS

1. **What is a metric space?**
   a. A set equipped with a distance function that satisfies specific properties.
   b. A set with no limit points.
   c. A set where every subset is open.
   d. A set where every function is continuous.
2. **Which of the following statements is true about continuity in metric spaces?**
   a. Every function is continuous in a metric space.
   b. A function is continuous if and only if it preserves convergent sequences.
   c. A function is discontinuous if it maps open sets to closed sets.
   d. Continuous functions in metric spaces are always differentiable.
3. **What does it mean for a sequence {xn} in a metric space (X, d) to converge to a point $x \in X$?**
   a. For every $\varepsilon > 0$, there exists an N such that $d(xn, x) < \varepsilon$ for all $n \geq N$.
   b. For every $\delta > 0$, there exists an N such that $d(xn, x) < \delta$ for all $n \leq N$.
   c. The sequence {xn} must have all its terms equal to x.
   d. Convergence of sequences does not apply in metric spaces.

4. **What defines a topology on a set X?**
   a. The set of all subsets of X.
   b. A collection of open sets satisfying certain properties.
   c. The set of all functions from X to itself.
   d. A collection of closed sets satisfying certain properties.
5. **How are topologies often constructed using bases?**
   a. By taking the union of all possible open sets.
   b. By defining a collection of subsets whose unions cover all open sets.
   c. By defining closed sets and then taking complements.
   d. By defining the set of all possible continuous functions.
6. **What role do neighborhood bases play in topology?**
   a. They determine the smallest possible open sets in a topology.
   b. They define neighborhoods that are maximal in size.
   c. They are used to define closed sets within a space.
   d. They provide a way to visualize topological structures.
7. **Which of the following statements is true about subspaces of topological spaces?**
   a. Subspaces have the same topology as their parent spaces.
   b. The closure of a subset in a subspace is always equal to the closure in the parent space.
   c. Interior points in a subspace are always interior points in the parent space.
   d. The boundary of a subset in a subspace is always equal to the boundary in the parent space.

REVIEW QUESTIONS

1. Define a metric space and explain its fundamental properties.
2. Discuss the concept of continuity in metric spaces. Provide examples to illustrate continuous and discontinuous functions in this context.
3. Explain how convergence is defined in metric spaces. Compare and contrast sequential and epsilon-delta definitions of convergence.
4. What are topologies? Describe the basic elements of a topology and provide examples of different types of topologies.
5. How are topological spaces constructed using bases and subbases? Give examples to demonstrate how bases can uniquely define a topology.
6. Discuss the role of neighborhood bases in defining local properties of points within a topological space.
7. Explain the concepts of closure, interior, boundary, and derived sets in the context of subspaces of topological spaces.

Answers to Multiple Questions

1. (a) 2. (b) 3. (a) 4. (b) 5. (b)

6. (a) 7. (c)

REFERENCES

1. Aish, R., Jabi, W., Lannon, S., Wardhana, N. M., & Chatzivasileiadi, A. (2018, September). Topologic: Tools to explore architectural topology. In *AAG* (Vol. 3, pp. 316–341).
2. Akiner, V. T. (1986). Topology-1: a knowledge-based system for reasoning about objects and spaces. *Design Studies*, *7*(2), 94–105.
3. Alcantud, J. C. R. (2020). Soft open bases and a novel construction of soft topologies from bases for topologies. *Mathematics*, *8*(5), 672.
4. Aleman, A., & Korenblum, B. (2008). Derivation-invariant subspaces of C∞. Computational Methods and Function Theory, 8(2), 493–512.
5. Arens, R. F. (1946). A topology for spaces of transformations. *Annals of Mathematics*, *47*(3), 480–495.
6. Arkhangel'ski, A. V., & Fedorchuk, V. V. (1990). The basic concepts and constructions of general topology. In *General Topology I: Basic Concepts and Constructions Dimension Theory*. Berlin, Heidelberg: Springer Berlin Heidelberg. (Vol. 2, pp. 1–90).
7. Banaschewski, B. (1964). Extensions of topological spaces. *Canadian Mathematical Bulletin*, *7*(1), 1–22.
8. Banon, G. J. F. (2000). New insight on digital topology. *Mathematical Morphology and its Applications to Image and Signal Processing*, 8, 139–148.
9. Bayoumi, F., & Rodabaugh, S. E. (2010). Overview and comparison of localic and fixed-basis topological products. *Fuzzy Sets and Systems*, *161*(18), 2397–2439.
10. Beer, G. (1985). On convergence of closed sets in a metric space and distance functions. *Bulletin of the Australian Mathematical Society*, *31*(3), 421–432.
11. Bhattacharya, S., Jhunjhunwala, A., Halder, A., Bhattacharyya, D., & Mitra, A. (2019). Going beyond base-pairs: topology-based characterization of base-multiplets in RNA. *RNA*, *25*(5), 573–589.
12. Bing, R. H. (1951). Metrization of topological spaces. *Canadian Journal of Mathematics*, *3*, 175–186.
13. Bonsangue, M. M., van Breugel, F., & Rutten, J. J. (1998). Generalized metric spaces: completion, topology, and powerdomains via the Yoneda embedding. *Theoretical Computer Science*, *193*(1–2), 1–51.
14. Bozkaya, T., & Ozsoyoglu, M. (1997, June). Distance-based indexing for high-dimensional metric spaces. In *Proceedings of the 1997 ACM SIGMOD International Conference on Management of Data* (Vol. 3, pp. 357–368).

15. Brennan, J. E. (1979). Point evaluations, invariant subspaces and approximation in the mean by polynomials. *Journal of Functional Analysis*, *34*(3), 407–420.
16. Brown, P. (1996). Comprehensive neighborhood-based initiatives. *Cityscape*, 2(3), 161–176.
17. Bukatin, M., Kopperman, R., Matthews, S., & Pajoohesh, H. (2009). Partial metric spaces. *The American Mathematical Monthly*, *116*(8), 708–718.
18. Burks, A. W. (1951). The logic of causal propositions. *Mind*, *60*(239), 363–382.
19. Çakallı, H., Sönmez, A., & Genç, Ç. (2012). On an equivalence of topological vector space valued cone metric spaces and metric spaces. *Applied Mathematics Letters*, *25*(3), 429–433.
20. Ceder, J. G. (1961). Some generalizations of metric spaces. *Pacific J. Math*, *11*(1), 105–125.
21. Choi, M. G., Ahn, S. J., Choi, J. H., Cho, S. M., & Yun, S. Y. (2020). Adaptive protection method of distribution networks using the sensitivity analysis for changed network topologies based on base network topology. *IEEE Access*, *8*, 148169–148180.
22. Clark, B., & Schneider, V. (1994). A comparison of topologies on free groups. *Kyungpook Mathematical Journal*, *34*(1), 127–135.
23. Dickman, R., & Porter, J. (1975). θ-closed subsets of Hausdorff spaces. *Pacific Journal of Mathematics*, *59*(2), 407–415.
24. Eriksson, D. M. (2003). A framework for the constitution of modelling processes: A proposition. *European Journal of Operational Research*, *145*(1), 202–215.
25. Fakcharoenphol, J., Rao, S., & Talwar, K. (2003, June). A tight bound on approximating arbitrary metrics by tree metrics. In *Proceedings of the Thirty-Fifth Annual ACM Symposium on Theory of Computing* (Vol. 2, pp. 448–455).
26. Ferenczi, V. (2005). Topological 0–1 laws for subspaces of a Banach space with a Schauder basis. *Illinois Journal of Mathematics*, *49*(3), 839–856.
27. Ferrando, J. C., & López-Pellicer, M. (2022). Classifying topologies through G-bases. *Axioms*, *11*(12), 744.
28. Flagg, B., & Kopperman, R. (1997). Continuity spaces: Reconciling domains and metric spaces. *Theoretical Computer Science*, *177*(1), 111–138.
29. Fomin, S. (1943). Extensions of topological spaces. *Annals of Mathematics*, *44*(3), 471–480.
30. Frank, M. J. (1971). Probabilistic topological spaces. *Journal of Mathematical Analysis and Applications*, *34*(1), 67–81.
31. Franklin, S. P., & Rajagopalan, M. (1971). Some examples in topology. *Transactions of the American Mathematical Society*, *155*(2), 305–314.
32. Fuster, F., & Silvi, B. (2000). Determination of protonation sites in bases from topological rules. *Chemical Physics*, *252*(3), 279–287.
33. Giuli, E. (1980). Bases of topological epi-reflections. *Topology and its Applications*, *11*(3), 265–273.

34. Gleason, A. M. (1958). Projective topological spaces. *Illinois Journal of Mathematics*, *2*(4A), 482–489.

35. Goffinet, D. (1991). Number systems with a complex base: a fractal tool for teaching topology. *The American Mathematical Monthly*, *98*(3), 249–255.

36. Gruenhage, G. (1984). Generalized metric spaces. In *Handbook of Set-Theoretic Topology*. North-Holland. (Vol. 3, pp. 423–501).

37. Guest, A. M., Cover, J. K., Matsueda, R. L., & Kubrin, C. E. (2006). Neighborhood context and neighboring ties. *City & community*, *5*(4), 363–385.

38. Guillemin, V., & Pollack, A. (2010). *Differential Topology*. American Mathematical Soc. (Vol. 370, pp. 76–80).

39. Halmos, P. R. (1969). Two subspaces. *Transactions of the American Mathematical Society*, *144*, 381–389.

40. Han, J., Zheng, K., Sun, A., Shang, S., & Wen, J. R. (2016, May). Discovering neighborhood pattern queries by sample answers in knowledge base. In *2016 IEEE 32nd International Conference on Data Engineering (ICDE)*. IEEE. (Vol. 3, pp. 1014–1025).

41. Imran, M., Baig, A. Q., Rashid, S., & Semaničová-Feňovčíková, A. (2018). On the metric dimension and diameter of circulant graphs with three jumps. *Discrete Mathematics, Algorithms and Applications*, *10*(01), 1850008.

42. Isaacs, R. R. (1948). The neighborhood theory: An analysis of its adequacy. *Journal of the American Institute of Planners*, *14*(2), 15–23.

43. Kahalon, O., Le, H., Milenković, L., & Solomon, S. (2022, July). Can't see the forest for the trees: Navigating metric spaces by bounded hop-diameter spanners. In *Proceedings of the 2022 ACM Symposium on Principles of Distributed Computing* (Vol. 3, pp. 151–162).

44. Kelley, J. L., Namioka, I., Donoghue, W. F., Lucas, K. R., Pettis, B. J., Poulsen, E. T., ... & Smith, K. T. (1963). *Linear topological spaces*. Springer Berlin Heidelberg. (Vol. 3, pp. 26–82).

45. Kent, D. C. (1966). On the order topology in a lattice. *Illinois Journal of Mathematics*, *10*(1), 90–96.

46. Ketterer, C. (2015). Cones over metric measure spaces and the maximal diameter theorem. *Journal de Mathématiques Pures et Appliquées*, *103*(5), 1228–1275.

47. Khayyeri, R., & Mohamadian, R. (2011). On base for generalized topological spaces. *Int. J. Contemp. Math. Sciences*, *6*(48), 2377–2383.

48. Kong, T. Y., Roscoe, A. W., & Rosenfeld, A. (1992). Concepts of digital topology. *Topology and its Applications*, *46*(3), 219–262.

49. Kowalkowski, C., Persson Ridell, O., Röndell, J. G., & Sörhammar, D. (2012). The co-creative practice of forming a value proposition. *Journal of Marketing Management*, *28*(13–14), 1553–1570.

50. Kremer, M. (1992). The multiplicity of general propositions. *Nous*, *26*(4), 409–426.

51. Kundu, S., Aggarwal, M., & Hazra, S. (2017). Finitely chainable and totally bounded metric spaces: Equivalent characterizations. *Topology and its Applications*, *216*, 59–73.

52. Levenshtein, V. I. (1992). Designs as maximum codes in polynomial metric spaces. *Acta Applicandae Mathematica, 29*, 1–82.

53. Levine, N. (1961). On the community of the closure and interior operators in topological spaces. *The American Mathematical Monthly, 68*(5), 474–477.

54. Li, L., & Li, Q. (2015). On enriched L-topologies: Base and subbase. *Journal of Intelligent & Fuzzy Systems, 28*(6), 2423–2432.

55. Lin, T. Y. (1988, February). Neighborhood systems and approximation in relational databases and knowledge bases. In *Proceedings of the 4th International Symposium on Methodologies of Intelligent Systems.* California State University Northridge CA. (Vol. 3, pp. 75–86).

56. Loeb, P. A. (2016). Topological spaces. In *Real Analysis.* Cham: Springer International Publishing. (Vol. 4, pp. 147–178).

57. Lógó, J., & Ismail, H. (2020). Milestones in the 150-year history of topology optimization: A review. *Computer Assisted Methods in Engineering and Science, 27*(2–3), 97–132.

58. Ma, W., Trusina, A., El-Samad, H., Lim, W. A., & Tang, C. (2009). Defining network topologies that can achieve biochemical adaptation. *Cell, 138*(4), 760–773.

59. MacKintosh, F. C., & Lubensky, T. C. (1991). Orientational order, topology, and vesicle shapes. *Physical review letters, 67*(9), 1169.

60. Mader, A. (2006, October). Basic concepts of functorial topologies. In *Abelian Group Theory: Proceedings of the Oberwolfach Conference, January 12–17, 1981*. Berlin, Heidelberg: Springer Berlin Heidelberg. (Vol. 3, pp. 251–271).

61. Manetti, M. (2023). *Topology.* Springer Nature. (Vol. 153, pp. 35–38).

62. McArthur, C. W., & Retherford, J. R. (1965). Uniform and equicontinuous Schauder bases of subspaces. *Canadian Journal of Mathematics, 17*, 207–212.

63. Meadows, S. O., Miller, L. L., Miles, J. N., Gonzalez, G. C., & Dues, B. T. (2013). Exploring the association between military base neighborhood characteristics and soldiers' and airmen's outcomes. *Rand Health Quarterly, 3*(1), 65–66.

64. Møller, J. M. (2007). General topology. *Matematisk Institut, Universitetsparken, 5*, 984–990.

65. Mormann, T. (2013). Topology as an Issue for History of Philosophy of Science. In *New challenges to philosophy of science.* Dordrecht: Springer Netherlands. (Vol. 3, pp. 423–434).

66. Mustafa, Z., & Sims, B. (2006). A new approach to generalized metric spaces. *Journal of Nonlinear and Convex Analysis, 7*(2), 289.

67. Naidu, S. V., Rao, K. P. R., & Rao, N. S. (2004). On the topology of D-metric spaces and generation of D-metric spaces from metric spaces. *International Journal of Mathematics and Mathematical Sciences, 2004*(51), 2719–2740.

68. Nicolau, M., Levine, A. J., & Carlsson, G. (2011). Topology based data analysis identifies a subgroup of breast cancers with a unique mutational profile and excellent survival. *Proceedings of the National Academy of Sciences, 108*(17), 7265–7270.

69. Pagliani, P., & Chakraborty, M. (Eds.). (2008). *A Geometry of Approximation: Rough Set Theory: Logic, Algebra and Topology of Conceptual Patterns*. Dordrecht: Springer Netherlands. (Vol. 3, pp. 21–23).

70. Pang, B., & Xiu, Z. Y. (2019). An axiomatic approach to bases and subbases in L-convex spaces and their applications. *Fuzzy Sets and Systems*, *369*, 40–56.

71. Phillips, J. W. (2013). On topology. *Theory, Culture & Society*, *30*(5), 122–152.

72. Pitkäranta, J. (1979). Boundary subspaces for the finite element method with Lagrange multipliers. *Numerische Mathematik*, *33*(3), 273–289.

73. Prager, S. D., & Reiners, W. A. (2009). Historical and emerging practices in ecological topology. *ecological complexity*, *6*(2), 160–171.

74. Robertson, A. P., & Robertson, W. (1980). *Topological vector spaces.* CUP Archive. (Vol. 53, pp. 98–100).

75. Rocha, P., & Caires, L. (2021). Propositions-as-types and shared state. *Proceedings of the ACM on Programming Languages*, *5*(1), 1–30.

76. Rodabaugh, S. E. (1999). Categorical foundations of variable-basis fuzzy topology. In *Mathematics of fuzzy sets: logic, topology, and measure theory.* Boston, MA: Springer US. (Vol. 3, pp. 273–388).

77. Rotman, B. (2012). Topology, algebra, diagrams. *Theory, Culture & Society*, *29*(4–5), 247–260.

78. Shabir, M., & Naz, M. (2011). On soft topological spaces. *Computers & Mathematics with Applications*, *61*(7), 1786–1799.

79. Shankar, R. (1977). Applications of topology to the study of ordered systems. *Journal de Physique*, *38*(11), 1405–1412.

80. Sharma, P. L. (1971). Proximity bases and subbases. *Pacific Journal of Mathematics*, *37*(2), 515–526.

81. Simon, C., Koniusz, P., Nock, R., & Harandi, M. (2020). Adaptive subspaces for few-shot learning. In *Proceedings of the IEEE/CVF conference on computer vision and pattern recognition* (Vol. 3, pp. 4136–4145).

82. Skriganov, M. M. (2017). Point distributions in compact metric spaces. *Mathematika*, *63*(3), 1152–1171.

83. Spreen, D. (1998). On effective topological spaces. *The Journal of Symbolic Logic*, *63*(1), 185–221.

84. Stadler, B. M., Stadler, P. F., Wagner, G. P., & Fontana, W. (2001). The topology of the possible: Formal spaces underlying patterns of evolutionary change. *Journal of theoretical biology*, *213*(2), 241–274.

85. Su, S. H., & Xu, Z. J. (2009). The connectedness relative to a subbase for the L-fuzzy topology. In *Fuzzy Information and Engineering Volume 2*. Springer Berlin Heidelberg. (Vol. 3, pp. 679–686).

86. Szymański, A., & Turzański, M. (1982). Subbase characterization of special spaces. In *Colloquium Mathematicum* (Vol. 2, pp. 185–197).

87. Varadarajan, V. S. (1965). Measures on topological spaces. *Amer. Math. Soc. Transl*, *48*(2), 161–228.

88. Von Neumann, J. (1935). On complete topological spaces. *Transactions of the American Mathematical Society*, *37*(1), 1–20.

89. Wallman, H. (1938). Lattices and topological spaces. *Annals of Mathematics*, *39*(1), 112–126.

90. Walter, H. (1975). Topologies on formal languages. *Mathematical systems theory*, *9*(2), 142–158.

91. Warren, R. H. (1978). Neighborhoods, bases and continuity in fuzzy topological spaces. *The Rocky Mountain Journal of Mathematics*, 4, 459–470.

92. Wicke, H. H., & Worrell Jr, J. M. (1975). Primitive structures in general topology. In *Studies in Topology*. Academic Press. (Vol. 3, pp. 581–599).

93. Wrigley, A. (2006). Abstracting propositions. *Synthese*, *151*, 157–176.

94. Zegura, E. W., Calvert, K. L., & Donahoo, M. J. (1997). A quantitative comparison of graph-based models for Internet topology. *IEEE/ACM Transactions on networking*, *5*(6), 770–783.

95. Zhu, P., Hu, Q., Han, Y., Zhang, C., & Du, Y. (2016). Combining neighborhood separable subspaces for classification via sparsity regularized optimization. *Information Sciences*, *370*, 270–287.

CHAPTER 2

Continuity and the Product Topology

LEARNING OBJECTIVES

After studying this chapter, you will be able to:

1. Understand the definition and fundamental properties of continuous functions in topological spaces.
2. Identify and apply equivalent conditions for continuity.
3. Utilize basis and subbasis concepts to explore continuity.
4. Analyze various examples of continuous functions to understand their applications and significance.
5. Comprehend the role of inclusion maps in relative topology.
6. Apply the Gluing Lemma to construct continuous functions.
7. Differentiate between open and closed maps, and understand their respective characterizations.
8. Explore the definition, properties, and applications of product topologies, including higher dimensions and geometric interpretations.

KEY TERMS FROM THIS CHAPTER

Continuous Functions
Basis and Subbasis
Relative Topology
Open Maps
Product Topology
Equivalent Conditions for Continuity
Inclusion Map
Gluing Lemma
Closed Maps

UNIT INTRODUCTION

The central notion in topology is the concept of *continuity of functions* between topological spaces. The first section of the chapter focuses on this concept and other related notions concerning mappings. In the second section, we address the problem of topologizing the Cartesian products of a family of topological spaces in a natural and meaningful way. Our primary focus is on the *Tychonoff topology,* which is the smallest topology on the product of topological spaces that ensures each projection map is continuous (Sgro, 1976).

2.1. CONTINUOUS FUNCTIONS

Here is an explanation of continuous functions (Figure 2.1).

2.1.1. Definition of Continuous Functions

In topology, the definition of a continuous function is analogous to its definition in analysis. This makes it straightforward to understand and apply.

Definition 2.1.1: Let (X) and (Y) be topological spaces. A function $(f: X \to Y)$ is called continuous if $f^{-1}(U)$ is open in (X \) for every open set $(U \subseteq Y)$ (Chittenden, 1929).

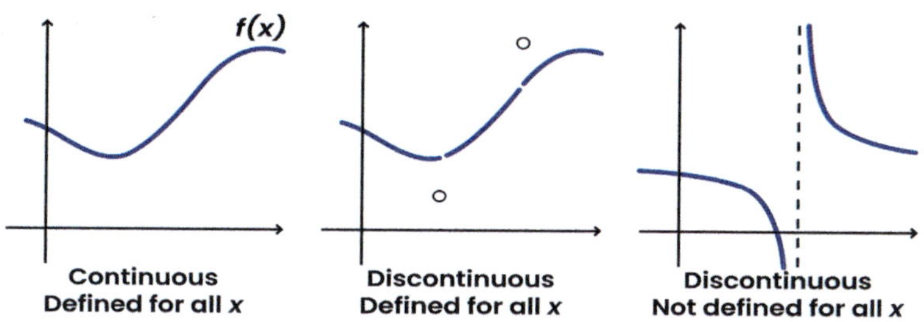

Figure 2.1. Illustration of continuous and discontinuous functions (Source: House of Math, Creative Commons License)

Remember

Remember the definitions and conditions for continuity in functions, especially in the context of topological spaces.

Examples:

1. Constant Function: A constant function $(c^{-1}(U))$ is continuous because $(c^{-1}(U))$ is either $(\emptyset)$ or (x) for every $(U \subseteq Y)$.
2. Discrete Space: Every function defined on a discrete space is continuous since every subset of a discrete space is open (Di Maio et al., 1998).

2.1.2. Equivalent Conditions for Continuity

Theorem 2.1.2: Consider (X) and (Y) as topological spaces and let $(f: X \to Y)$ be a function. The following criteria are equivalent (Davoren, J1999).

i. (f) is continuous.

ii. $(f^{-1}(F))$ is closed in (X) for every closed set $(F \trianglelefteq Y)$.

iii. $\left(f(\bar{A}) \subseteq \overline{f(A)}\right)$ for every set $(A \subseteq X)$.

iv. $\left(f^{-1}(\bar{B}) \subseteq \overline{f^{-1}(B)}\right)$ for every set $(B \trianglelefteq Y)$.

Proof:

i. $(a) \Leftrightarrow (b)$: This follows from the equality $\left(f^{-1}(Y \setminus B) = X \setminus f^{-1}(B)\right)$.

ii. $(b) \Rightarrow (c)$: $If \left(f(\bar{A}) \subseteq \overline{f(A)}\right), then \left(\bar{A} \subseteq f^{-1}\left(\overline{f(A)}\right)\right)$, which implies

$\left(f^{-1}(\bar{F}) = \overline{f^{-1}(F)}\right) for closed (F)$.

iii. $(c) \Rightarrow (d)$: Taking $(A = f \wedge \{-1\}(B))\ in (c)$, we get $\left(f\left(\overline{f^{-1}(B)}\right) \subseteq \overline{f\left(f^{-1}(B)\right)} \subseteq \bar{B}\right)$, so $\left(f^{-1}(\bar{B}) \subseteq \overline{f^{-1}(B)}\right)$.

iv. $(d) \Rightarrow (b)$: $If (F)$ is closed in (Y), then $\left(f^{-1}(F) \subseteq \overline{f^{-1}(F)}\right)$, which is always true. Therefore, $(f^{-1}(F))$ is closed (Kowalczyk & Turowska, 2021).

2.1.3. Basis and Subbasis for Continuity

Theorem 2.1.3: Let (X) and (Y) be topological spaces. A function $(f: X \to Y)$ is continuous if and only if the inverse image of every set in a basis (or subbasis) of (Y) is open.

Proof:

i. If (B) is a basis for (Y), then for (f) to be continuous, $(f^{-1}(B_i))$ must be open in (X) for every (B_i) in the basis.

ii. If (S) is a subbasis for (Y), and $(f^{-1}(S))$ is open in (X) for every $(S \in S)$, then for any open set $U \subseteq Y),(f^{-1}(U) is$ open in (X).

2.1.4. Examples of Continuous Functions

Example 2.1.3: The function $(f:(0,2\pi) \rightarrow S^1)$ defined by $f(t)=e^{it}$ is continuous. The open arcs of the circle (S^1) form a basis for its topology. For an open arc (G) not containing $1 \in S^1, f^{-1}(G)$ is an open interval in $(0,2\pi)$ If (G) contains (1), then $f^{-1}(G)$ is the union of two open intervals in $(0,2\pi)$ (Ciesielski & Rosłanowski, 2000) (Figure 2.2).

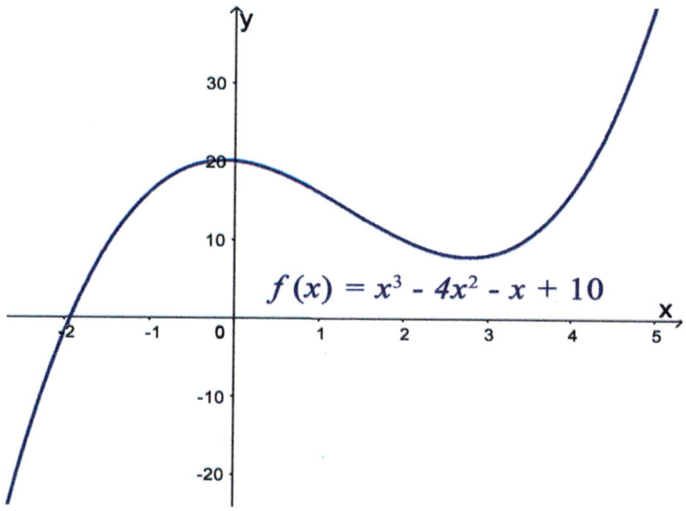

Figure 2.2. Example of a continuous function (Source: Story of Mathematics, Creative Commons License)

2.1.5. Continuity of Identity and Composition

Theorem 2.1.4:

i. The identity map $(1_X : X \rightarrow X)$ is continuous.

ii. If $(f : X \rightarrow Y)$ and $(g : Y \rightarrow Z)$ are continuous, then the composition $(g \circ f : X \rightarrow Z)$ is continuous (Elliott et al., 2023).

2.1.6. Inclusion Map and Relative Topology

For a subspace $(A \subseteq X)$, the inclusion $map(j : A \to X)$ is continuous in the relative topology. The relative topology (T_A) on (A) is the smallest topology making the inclusion map continuous.

If $(f : X \to Y)$ is continuous, then the restriction $(f|_A : A \to Y)$ is continuous. If $(f(X) \subseteq Y)$ has the relative topology, then $f : X \to f(X)$ is continuous (Westerveld & Knowles, 2021) (Figure 2.3).

Did you know?

Did you know that the inclusion map between a subspace and the whole space induces a topology on the subspace, known as the relative topology?

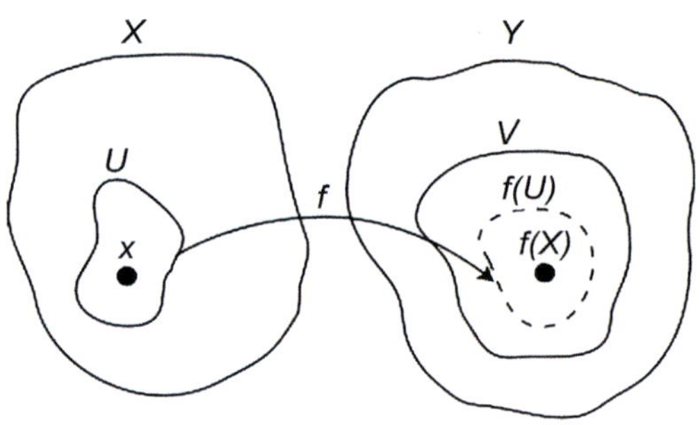

Figure 2.3. Topological formulation of continuity of a function at a point described on the page continuity (topology) (Source: Derrick Coetzee, Creative Commons License)

2.1.7. Localized Continuity

Definition 2.1.5 (Cauchy): A function $(f : X \to Y)$ is continuous at $(x \in X)$ if, for every neighborhood $(N) of (f(x))$ in (Y), there exists a neighborhood (M) of (x) in (X) such that $f(M) \subseteq N$.

Theorem 2.1.6: A function $(f : X \to Y)$ is continuous if and only if it is continuous at each point of (X) (Forstneric & Rosay, 1987).

Proof:

i. If (f) is continuous, then for any neighborhood (N) of $f(x)), (f^{-1}(N)$ is a neighborhood of (x), proving continuity at (x).

ii. If (f) is continuous at each point, for any open set $(V \subseteq Y), (f^{-1}(V))$ is a neighborhood of each $x \in f^{-1}(V)$, making $(f^{-1}(V))$ open in (X), hence (f) is continuous.

Continuity of Addition and Multiplication in (R)

i. Addition Function $(\alpha(x,y)=x+y)$: The addition function is continuous in the Euclidean topology induced by the metric $(d^*(p,q)=\max(|x-s|,|y-t|))$.

ii. Multiplication Function $(\mu(x,y)=xy)$: The multiplication function is continuous, as shown by considering neighborhoods and their images under (μ) (Ajmal & Sarma, 1992).

2.1.8. Gluing Lemma

To construct a continuous function on a set (X) covered by subsets $\{A_i\}$, where $(X = \cup A_i)$and $(f_i : A_i \rightarrow Y)$ are continuous with $(f_i = f_j)$on $(A_i \cap A_j)$, there exists a unique continuous function $(f : X \rightarrow Y)$ such that $(f|_{A_i} = f_i)$ for all (i). This is known as the Gluing (or Pasting) Lemma (Dimofte & Garoufalidis, 2013) (Figure 2.4).

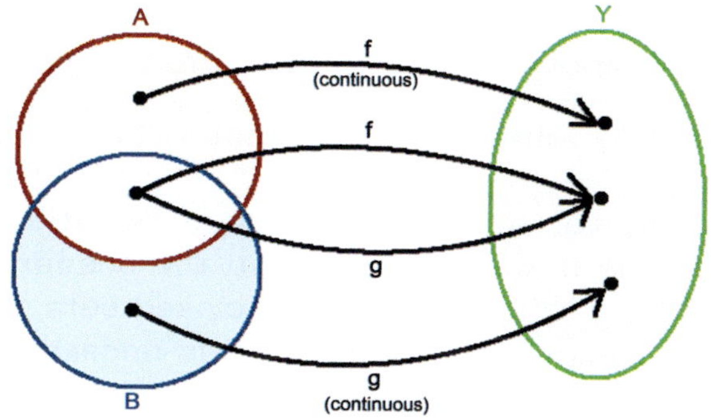

Figure 2.4. To show that h is continuous we will show that for every closed set V⊆Y we have that $h^{-1}(V)$ is closed in X (Source: Math Online, Creative Commons License)

Lemma 2.1.7. Consider (X) and (Y) as topological spaces, and let $(A_1, \ldots, A_n)$ be a finite set of continuous functions for every $(i=1,\ldots,n)$. If $f|_{A_i \cap A_j} = f|_{A_i \cap A_j}$) for all (i) and (j), then the function (f)defined by $(f_{Ai} = f_i)$for all (i) is continuous (Kamae et al., 2015).

Proof:

Let's assume that each (A_i) is closed in(X), and consider $(F \subseteq Y)$ as an arbitrary closed set. By the continuity of (f_i), the inverse image $(f_i),(f_i^{-1}(F))$ is a closed set in(A_i) (Hadjigeorgiou, 1999). Given that (A_i) is closed in (X) for every (i), it follows that each $(f_i^{-1}(F))$ is closed in (X). Hence, the set $(f^{-1}(F) = \cup_{i=1}^{n} f_i^{-1}(F))$ is closed in (X). Similar reasoning applies to the case where the $(A_i)'s$ are open. It is obvious that the lemma above applies to any family of open sets (A_i). However, one-point sets (x) in an interval $((a,b) \subseteq R)$.readily demonstrate that this is not achievable in the scenario involving infinite families of closed sets. In order to solve the problem, it is necessary to impose specific restrictions on the placement of the closed sets (A_i)in (X) (Garoufalidis et al., 2015).

Definition 2.1.8: In (X), every point has a neighborhood $(U)$where the intersection of $(U \cap A_i = \varnothing)$ for only a finite number of indices (i). This collection of subsets, denoted by $(\{A_i\})$, in a given space $(X)i$ s referred to as locally finite (or nbd-finite) (Dedushenko, 2020).

Proposition 2.1.9: Consider subsets family $(\{A_i\})$ of a locally finite space (X).

Proof: For every point $(x \in X)$, there exists an open neighborhood(U) such that the intersection of $(U \cap A_i = \varnothing)$is empty for only a limited number of indices (i). Similarly, the sets $(U \cap A)$, also follow this property, as $(U \cap A \subseteq X - A_i)$ is an open set, implying that (A_i) is closed.

From the above principle, it is evident that the combination of a locally finite collection of closed sets forms a closed set (Bunke, 1995).

Corollary 2.1.10: A space (X) is considered to be the combination of a locally finite collection $(\{A_i\})$ of closed sets. In this scenario, a function (f)mapping from $(X)$to another space (Y) is considered continuous if and only if the function(f) restricted to each (A_i) is also continuous.

Proof: Assume that for each(i), $(f|_{A_i})$ is continuous. $(f^{-1}(F) = \cup_i f^{-1}(F) \cap A_i)$is closed in (X), if $(f)$is a closed subset of (Y),. As a result, $(f)$is continuous. The opposite is evident. When defining a continuous function $(f : X \rightarrow Y)$ we often begin by splitting \$(X)$ nto closed (or open) subsets \$(A_i)$\) and defining (f) on each of these subsets independently in a way that makes $(f|_{A_i})$clearly continuous and ensures that the various definitions concur on the overlaps (Baraglia & Konno, 2020).

Now, we will examine the concept of "equivalence" in relation to topological spaces.

Definition 2.1.11: A homeomorphism is a bijective function $(f : X \rightarrow Y)$between spaces (X) and (Y) that is continuous, as well as its inverse function (f^{-1}). Two spaces (X) and $(Y)$are considered homeomorphic, represented by $(X \approx Y)$, if there exists a homeomorphism $(X \rightarrow Y)$.

It is obvious that a continuous function $(f : X \to Y)$ is a homeomorphism if and only if $(g : Y \to X)$ exists, such that $(f \circ g = 1_Y)$ and $(g \circ f = 1_X)$. A continuous bijection is not always a homeomorphism, it should be noted. Any element in a discrete topological set can be mapped to an identical element in the same set with a different topology using a function called the identity map (Bolsinov & Matveev, 2011).

Example 2.1.4: The open ball $(B(0;1) = \{x \in R^n \mid |x| < 1\})$ is homeomorphic to the Euclidean space (R^n). A homeomorphism of $(x \mapsto x/(1+|x|))$ is represented by the map $(x \mapsto x/(1+|x|))$, whose inverse is $(x \mapsto x/(1-|x|))$.

Example 2.1.5: The punctured sphere $(S^n - \{p\})$, is homeomorphic to $(\backslash mathbb\{R\}^\wedge n \backslash)$, where $(p = (0,\ldots,0,1))$. As a matter of fact, the mapping $(f : S^n - \{p\} \to R^n)$ defined by \ $(f(x_0,\ldots,x_n) = (x_0,\ldots,x_{n-1})/(1-x_n))$ is easily shown to be a homeomorphism, and the inverse mapping is provided by $(y \mapsto (y_0,\ldots,y_{n-1},1-|y|^2))$, where $(y = (y_0,\ldots,y_{n-1}))$ and $(|y|^2 = \sum_{i=0}^{n-1} y_i^2)$. The stereographic projection is the name given to the homeomorphism (f).

Example 2.1.6: Observe that a bijection $(C^n \approx R^{2n})$ is defined by the correspondence $(a+ib \mapsto (a,b))$ between (C) and (R^2). This bijection is obviously a distance-preserving map. Consequently, the Euclidean space is homeomorphic to the unitary space $(\dashv o\{C\}^\wedge n)$. Similar to this, a homeomorphism is the bijection $(q \mapsto (x_0,x_1,x_2,x_3))$ where (q) is the quaternion $(x_0 + x_1 i + x_2 j + x_3 k)$. This correspondence defines the Euclidean space (R^{4n}).

The standard approach to demonstrating the homeomorphism between two spaces involves establishing a pair of mutually inverse continuous functions connecting the two spaces. Regrettably, the creation of such functions is typically challenging. Below, we examine several characterizations of homeomorphisms, although these do not significantly simplify our task. In order to achieve this objective, we initially categorize functions between topological spaces into two distinct categories (Sénizergues, 2019).

Definition 2.1.12: A function $(f : X \to Y)$ is considered open if the image of every open set in (X) is also open in (Y). Likewise, a function is said to be closed if each closed set in (X) has an image that is also closed in (Y).

Example 2.1.7: Every function $(f : X \to Y)$ is obviously closed and open if (Y) is a discrete space.

Example 2.1.8: The map that includes (I) into (R) is closed but not open, unlike the map that includes $((0,2) \subseteq R)$, which is both open and closed.

Example 2.1.9: There is no closure or openness in the continuous function $(t \mapsto e^{it})$ of $([0,2\pi))$ onto (S^1). Even though its mapping image is closed, the interval $([0,\pi))$ is open in $([0,2\pi))$). Moreover, the mapping does not provide a closed image of interval $([\pi,2\pi))$ even if it is closed in $([0,2\pi))$.

The examples demonstrate the independence of continuous, open, and closure functions. In certain situations, demonstrating a map's closure from one topological space to another is simpler than demonstrating its inverse continuity. Thus, following homeomorphism explanation is important (Jaffe et al., 2011).

Theorem 2.1.13: Consider the function $(f : X \to Y)$, which is both bijective and continuous. A function (f) is a homeomorphism if and only if it is both open and closed.

Proof: For any subset $(U \subseteq X)$, we have $(f^{-1}(f(U)) = U)$.It is evident that the function (f^{-1}) is continuous if and only if the function $(f)$is open. Regarding the second equivalence, it can be observed that (f(X-U)=Y-f(U)) due to the bijectivity of (f). It is evident that $(f)$is open if and only if $(f)$is closed.

The previous theorem states that a homeomorphism $(h : X \to Y)$ produces a one-to-one relationship between the open sets (or closed sets) in $(X)$and (Y), as well as a one-to-one relationship between the points of (X) and (Y). Thus, any claim about (X) as a topological space likewise holds for any homeomorphism of (X). Due to this logic, spaces that are homeomorphic are referred to as topologically equivalent.

A topological property, or topological invariant, is a property of spaces that is shared by all homeomorphs of a space that has this characteristic. The study of topological characteristics that hold steady during repeated transformations is known as topology. The book will provide numerous examples of these invariants as we read it.

These invariants are frequently used to determine the homeomorphism between two provided spaces. Our study reveals that these spaces are not homeomorphic since they yield different answers when calculating a meaningful invariant for each of them. Not every property in a metric space is topologically invariant. Consider the concept of boundedness, as an example (Huang, 2011).

2.1.9. Closed Maps and Closure Operation

Proposition 2.1.14: Let (X) and (Y) be topological spaces and let $(f:X \to Y)$ be a function. Then (f) is a closed map if and only if $\left(f(\bar{A}) = \overline{f(A)}\right)$ for every subset $(A \subseteq X)$.

Proof:

(⇒): Assume (f) is a closed map. For any subset $(A \subseteq X)$, we have:

$$f(A) \subseteq f(\bar{A}) \subseteq \overline{f(A)}$$

Since (f) is closed, $(f(\bar{A}))$ is closed, and hence $\left(\overline{f(A)} \subseteq f(\bar{A})\right)$. Thus, $\left(f(\bar{A}) = \overline{f(A)}\right)$.

(⇐): Assume $\left(f(\bar{A}) = \overline{f(A)}\right)$ for every subset $(A \subseteq X)$. Let (A) be a closed subset of (X). Then $(\bar{A} = A)$, and we have:

$$f(A) = f(\bar{A}) = \overline{f(A)}$$

Thus, $(f(A))$ is closed in (Y), proving (f) is a closed map (Schmidt & Berghammer, 2011):

2.1.10. Continuous and Closed Functions

Corollary 2.1.15: A function $(f:X \to Y)$ between topological spaces is continuous and closed if and only if $\left(f(\bar{A}) = \overline{f(A)}\right)$ for every subset $(A \subseteq X)$.

This corollary implies that a homeomorphism is a bijective map that commutes with the closure operation.

(Long & Herrington, 1982).

2.1.11. Characterizations of Open Maps

Theorem 2.1.16: There are several equivalent properties of a function (f: X →Y \):

i. (a) (f) is an open map.

ii. (b) For every $(x \in X)$ and each neighborhood (U) of (x) in (X), there is a neighborhood (W) of $(f(x))$ in (Y) with $(W \subseteq f(U))$.

iii. (c) For every $(x \in X)$ and each basic neighborhood (B) of (x) in (X),$(f(B))$ is a neighborhood of $(f(x))$ in (Y).

iv. (d) The image of each member of a basis for (X) is open in (Y).

Proof: The proof is quite simple and requires demonstrating that each condition leads to the next (Baker, 1986). The reader is expected to work through the detailed proof on their own (Sayre et al., 2018).

2.1.12. Topological Embedding

Definition 2.1.17: An embedding, also known as a topological embedding, is a function $(f:X \to Y)$ that maps the space (X) onto a subspace $(f(X))$ of the space (Y) in a way that preserves the topological properties of (X).

i. The inclusion of a subspace in its ambient space is an embedding.

ii. A closed (or open) continuous injection is also an embedding.

iii. However, a continuous injection may not necessarily be an embedding (Bonner et al., 2017).

Example 2.1.10: The function $\left(f:(0,2\pi)\to R^2\right)$ defined by $\left(f(t)=e^{it}\right)$ is a continuous injective mapping. The set of values that (f) can take is the unit circle, denoted by $\left(S^1\right)$. Nevertheless (Liu et al., 2019), (f) cannot be considered an embedding since it does not successfully map open neighborhoods of (0) to open neighborhoods of (1).

2.2. PRODUCT SPACES

A basic creation in topology that extends the concept of topology beyond individual spaces to their products is the topology of product on the Cartesian product (X × Y) of two topological spaces (X) and (Y). As a result, we can investigate the characteristics of composite systems or spaces whose constituents contain both X and Y components (Grabowski, 1999).

2.2.1. Definition and Basis of Product Topology

The product topology on (X × Y) is defined by specifying a basis of open sets that respects the structure of (X) and (Y).

Tip:

The product topology on a Cartesian product of spaces is defined by the basis consisting of all products of open sets from each space.

Basic Idea: Given two topological spaces (X) and (Y), the product topology on (X × Y) is generated by sets of the form (U × V) where $(U \subseteq X)$ and $(V \subseteq Y)$ are open sets in (X) and (Y), respectively.

Basis Definition: The collection $\left(\mathcal{B} = \{U \times V \mid U \subseteq X \text{ open}, V \subseteq Y \text{ open}\}\right)$ forms a basis for the product topology on $(X \times Y)$.

Topology Generation: The product topology is the smallest topology on $(X \times Y)$ that makes the projections $(p_X : X \times Y \to X)$ and $(p_Y : X \times Y \to Y)$, defined by $(p_X(x, y) = x)$ and $(p_Y(x, y) = y)$, continuous.

2.2.2. Properties of the Product Topology

2.2.2.1. Continuity of Projections

The projection maps (p_X) and (p_Y) are continuous in the product topology. This means that the preimages of open sets under these projections are open in (X × Y). Formally, for any open set $(U \subseteq X)$ and $(V \subseteq Y), (p_X^{-1}(U) = U \times Y)$ and $(p_Y^{-1}(V) = X \times V)$, which are open in (X × Y)

2.2.2.2. Minimality

The product topology is defined as the coarsest topology on (X × Y) such that both (p_X) and (p_Y) are continuous. Any topology that is finer than this product topology and ensures the continuity of (p_X) and (p_Y) must include the basis $(\mathcal{B})$. Thus, the product topology is the smallest topology that guarantees the continuity of these projections (Comfort & Grant, 1981).

2.2.2.3. Topology Structure

The product topology is closed under finite intersections, meaning $((U_1 \times V_1) \cap (U_2 \times V_2) = (U_1 \cap U_2) \times (V_1 \cap V_2))$ is open in $(X \times Y)$.

However, it is not closed under arbitrary unions of open sets, which is why we use the basis $(\mathcal{B})$ to define it (Preuss, 2012).

2.2.2.4. Constructing the Basis

Using Basis from (X) and (Y):

i. If $(\mathcal{C})$ and $(\mathcal{D})$ are bases for (X) and (Y), respectively, then $(\{C \times D \mid C \in \mathcal{C}, D \in \mathcal{D}\})$ forms a basis for the product topology on $(X \times Y)$.

ii. This basis ensures that any open set in $(X \times Y)$ can be expressed as a union of these basis elements.

2.2.2.5. Examples and Applications

Euclidean Space (R^2):

i. In (R^2), the product topology coincides with the Euclidean topology. Here, open rectangles $((a,b) \times (c,d))$ form a basis for the product topology.

ii. This topology allows us to extend concepts like continuity and convergence from single-variable functions to functions of two variables (Rounds, 1984).

2.2.3. Higher Dimensions and Geometric Interpretations

2.2.3.1. Cylinders and Tori

The product topology helps define structures like cylinders $(S^1 \times I)$ and tori $(S^1 \times S^1)$. For example, $(S^1 \times I)$ represents a cylinder in (R^3), where (S^1) is the unit circle and (I) is the unit interval.

$(S^1 \times S^1)$ is the torus, resembling a doughnut shape. This product space captures the surface of revolution obtained by rotating a circle about an axis in (R^3)(Núñez et al., 2022) (Figure 2.5).

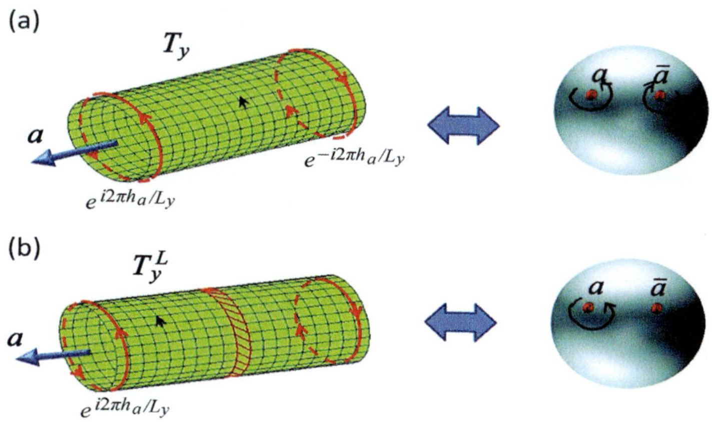

Figure 2.5. (a) A global translation of the cylinder is equivalent to spinning two quasiparticles on a sphere to opposite directions. (b) The partial translation T_y^L of the left part of the cylinder. In both panels, the red circles with arrows indicate the chiral edge states (Source: Hong-Hao Tu, Creative Commons License)

2.2.3.2. Metric Spaces

In metric spaces such as (R^n), the product topology produced by the Cartesian metric $d((x_1, \ldots, x_n), (y_1, \ldots, y_n)) = \sqrt{(x_1 - y_1)^2 + \cdots + (x_n - y_n)^2}$ is equivalent to the Euclidean topology.

This alignment guarantees that open balls in (R^n) are likewise open in the product topology, showing the coherence between multiple topological concepts (Bonsangue et al., 1998).

2.2.4. Associativity and Commutativity

2.2.4.1. Properties of Products

The product topology demonstrates both the associative and commutative features. For example, the expression $((X \times Y) \times Z)$ is homeomorphic to $(X \times (Y \times Z))$, which means that we can create Cartesian products of many spaces using an inductive approach.

2.2.4.2. Relativization

When examining subspaces $(A \subseteq X)$ and $(B \subseteq Y)$, it is worth noting that the product topology on (A × B) aligns with the relative topology induced from (X × Y) This offers a reliable framework for examining products of subspaces (Bhardwaj et al., 2020).

2.2.5. Importance in Mathematics

Following is the importance in mathematics.

2.2.5.1. Foundational Concept

Product topology is fundamental in the field of topology and mathematics as a whole. Mathematicians can use it to analyze complex formations created by combining more basic ones, serving as a connection between abstract spaces and geometric intuition.

2.2.5.2. Applications in Analysis and Geometry

- The product topology is useful in the analysis of multivariable functions and their features, such as convergence, continuity, and differentiability.
- Geometry utilizes the concept of higher-dimensional shapes and surfaces that are created by combining lower-dimensional components, helping in their definition and comprehension.

The product topology on (X ×Y) is a significant construction in topology, which establishes a topology that preserves the characteristics of both (X) and (Y) while enabling the examination of their collective features. It guarantees the uninterrupted flow of projections and serves as a foundation for examining places and buildings with more dimensions. The product topology is essential for generalizing topological concepts from Euclidean spaces to abstract topological spaces. It provides a coherent framework for understanding complicated mathematical objects by considering Cartesian products.

Through an examination of its definition, properties, examples, and applications, one can develop a greater understanding of the significance of the product topology in various areas of contemporary mathematics. This makes it a fundamental component of topological theory and practice, with connections to fields such as algebraic topology and differential geometry (Dubrovin et al., 2012).

2.2.6. Infinite Products and Product Topologies

The idea of product topologies in topology encompasses the extension of finite Cartesian products of topological spaces to include infinite Cartesian products. This extension is essential for comprehending and examining the characteristics that arise from infinite collections of spaces, guaranteeing that fundamental theorems from finite products remain valid in an infinite setting (Comfort & Negrepontis, 1972).

2.2.6.1. Definition and Construction

Cartesian Product: Given a family of topological spaces $(\{X_\alpha\}_{\alpha\in A})$, the Cartesian product $\left(\prod_{\alpha\in A} X_\alpha\right)$ consists of all functions $(x: A \to \bigcup_{\alpha\in A} X_\alpha)$ such $that\,(x(\alpha) \in X_\alpha)$ for each $(\alpha \in A)$.

Box Topology: The box topology $on\left(\prod_{\alpha\in A} X_\alpha\right)$ is generated by sets of the form $\left(\prod_{\alpha\in A} U_\alpha\right)$, where each (U_α) is open in (X_α). This topology ensures that each projection $\left(p_\beta : \prod_{\alpha\in A} X_\alpha \to X_\beta\right)$, defined by $(p_\beta(x) = x(\beta))$, is continuous.

Product Topology (Tychonoff Topology): The product topology on $\left(\prod_{\alpha\in A} X_\alpha\right)$, also known as the Tychonoff topology, is the smallest topology that makes all the projections (p_α) continuous. It is generated by the subbasis $\left(\{p_\alpha^{-1}(U_\alpha) \mid U_{\alpha \text{ open in}} X_\alpha\}\right)$ (Torkestani, 2013).

2.2.6.2. Properties and Comparisons

Comparison with Box Topology: Although the box topology can be easily defined and allows arbitrary unions of open sets, it sometimes results in an excessive number of open sets. This can confuse theorems concerning continuity and convergence in infinite products.

Compactness and Tychonoff's Theorem: Tychonoff's theorem states that the product $\left(\prod_{\alpha\in A} X_\alpha\right)$ is compact if and only if each (X_α) is compact, as guaranteed by the Tychonoff topology. This theorem is a potent instrument for establishing properties of infinite product spaces.

Applications in Functional Analysis: The product topology in functional analysis enables the examination of infinite sequences or families of functions, assuring the preservation of limits and continuity qualities over numerous dimensions or variables (Niu, 2022).

2.2.6.3. Examples and Illustrations

Example 1: Real Line $\left(R^N\right)$:

Let's consider $\left(R^N\right)$, which is the set of all sequences of real numbers. The product topology on $\left(R^N\right)$, guarantees that sequences converge pointwise if and only if they converge in the product topology.

Example 2: Hilbert Space $\left(1^2\right)$:

Given sequences of complex numbers with a finite sum of squares, the Hilbert space $\left(1^2\right)$, has the product topology formed from $\left(C^N\right)$.In functional analysis, this topology makes it easier to examine spectral properties and orthogonal projections.

Example 3: Product of Compact Spaces:

If $(\{X_n\}_{n\in N})$ is a sequence of compact spaces, then the product space $(\prod_{n\in N} X_n)$. The product topology of the given set is also proven to be compact according to Tychonoff's theorem. The concept of compactness plays a crucial role in various branches of mathematics, such as topology and measure theory (Gallagher & Engelke Infante, 2023).

2.2.6.4. Importance in Mathematics

Topological Vector Spaces: The product topology is essential for establishing topologies on vector spaces of functions or sequences, which allows for the examination of continuity, convergence, and completeness in spaces with an infinite number of dimensions.

Algebraic Structures: The product topology is a useful tool in algebraic topology and category theory. It allows for the examination of products and coproducts of spaces and objects, enabling a deeper comprehension of universal qualities and homotopy classes.

Infinite Combinatorial Structures: The product topology is a useful tool in combinatorial mathematics for studying and understanding infinite collections of discrete structures, such as graphs or lattices. It ensures that important qualities, such as connectedness or order-preserving maps, are maintained (Rashevsky, 1954).

2.2.6.5. Challenges and Considerations

Topology of Pointwise Convergence: When considering function spaces, it is crucial to select the appropriate topology based on the specific characteristics of the functions under investigation, as the topology of pointwise convergence frequently varies from the product topology.

Measurable Spaces: When studying integrable functions and measures on product spaces in measure theory, the decision between using the product topology and the weaker topology of virtually everywhere convergence has a significant impact.

By expanding the definition of topology from finite to infinite collections, the product topology on Cartesian products of topological spaces offers a strong foundation for investigating characteristics in a variety of dimensions or variables. Compactness theorems, functional analysis, algebraic structures, and other applications rely on product topology, a fundamental component of contemporary mathematics that provides greater understanding of the behavior and structure of infinite-dimensional spaces.

Mathematicians can apply the product topology to a variety of research topics, guaranteeing continuity and coherence in the study of complex systems and structures in mathematics, by being aware of its definition, properties, examples, and applications. As a result, product topology serves as a fundamental concept in topology and related domains, assisting in the theoretical and analytical bridge between finite and infinite dimensions in mathematics (Habib, 2008).

This is another method of defining the product topology:

Proposition 2.2.7: When considering a collection of spaces $(\{X_\alpha\}_{\alpha\in A})$, the product topology on the Cartesian product of these spaces $\left(\prod_{\alpha\in A} X_\alpha\right)$ is defined as the most minimal topology that ensures all the projections $(p_\beta : \prod_{\alpha\in A} X_\alpha \to X_\beta)$. from the product space to each individual space are continuous.

Notice that, when considering $(U_\beta \subseteq X_\beta)$, the inverse image $(p_\beta^{-1}(U_\beta)$ can be defined as the set of elements $\left\{x \in \prod_{\alpha\in A} X_\alpha \mid x_\beta \in U_\beta\right\}$). In other words, $(p_\beta^{-1}(U_\beta))$ corresponds to the product $\left(\prod_{\alpha\in A} U_\alpha\right)$. where $(U_\alpha = X_\alpha)$ for $(\alpha \neq \beta)$.

The basis for the product topology on $\left(\prod_{\alpha\in A} X_\alpha\right)$ is made up of sets $\left(\prod_{\alpha\in A} U_\alpha\right)$. where $(U_\alpha \subseteq X_\alpha)$ are open and $(U_\alpha = X_\alpha)$ for all except a finite number of (α)'s.

Notably, all basic open sets in the product topology $\left(\prod_{\alpha\in A} X_\alpha\right)$ is also open in a box topology. The box topology is seen as more complex than the product topology. The product topology base on $\left(\prod_{\alpha\in A} X_\alpha\right)$ coincides with the box topology when the indexing set (A) is finite. As a result, the two topologies for $\left(\prod_{\alpha\in A} X_\alpha\right)$ coincide. Thus, given $(A = \{\alpha_1, \ldots, \alpha_n\})$, it is evident that the product space $\left(\prod_{\alpha\in A} X_\alpha\right)$ is naturally homeomorphic to the space $(X_{\alpha_1} \times \cdots \times X_{\alpha_n})$ that was previously defined. This homeomorphism is established by the map $(\eta : x \mapsto (x_{\alpha_1}, \ldots, x_{\alpha_n}))$. As a result, these two spaces are recognized as being the same.

Clearly, if (G) is a nonempty open subset of $\left(\prod_{\alpha\in A} X_\alpha\right)$, then it contains a nonempty basic open set.

Consequently, $(p_\alpha(G)=X_\alpha)$ for all but a finite number of (α). As a result, the product $\left(\prod_{\alpha\in A} U_\alpha\right)$, where each (U_α) represents a proper nonempty open set in (X_α), does not exhibit openness in the product topology, unless the family $(\{X_\alpha\}_{\alpha\in A})$ is finite. However, when considering closed sets product $(F_\alpha \subseteq X_\alpha)$ is denoted as $\left(\prod_{\alpha\in A} F_\alpha\right)$, it is important to note that this product is closed in the product space $\left(\prod_{\alpha\in A} X_\alpha\right)$. This is due to the fact that $(p_\alpha^{-1}(F_\alpha) = F_\alpha)$

To understand this, it is crucial to notice that every element of (S) is considered open in product space $\left(\prod_{\alpha\in A} X_\alpha\right)$. In the context of a finite collection $(S_1, \ldots, S_n) in (S_\alpha)$, where $(x \in \bigcap_{i=1}^{n} p_\alpha^{-1}(S_i) \subseteq p_\alpha^{-1}(\bigcup_{i=1}^{n} S_i))$. The inclusion property holds for the preimage of the union of sets under the inverse image of a function. Given that $(x \in p_\alpha^{-1}(U_\alpha))$ for any arbitrary value of x, we can conclude that $(p_\alpha^{-1}(U_\alpha))$ is open in the topology generated by the family (S). It can be concluded that the two topologies for the product $\left(\prod_{\alpha\in A} X_\alpha\right)$ are the same. This is because the product topology is generated by the sets $(p\_\alpha\%\%\%\%\{-1\}(U\_\alpha))$, where U_α represents the open subsets of (X_α) and ( )varies over (A).

Similarly, each (X_α) topology is determined by a base, the topology of product on $\left(\prod_{\alpha\in A} X_\alpha\right)$ can then be described by coordinate spaces basic open subsets (Alimi et al., 2022).

Proposition 2.2.8: Consider collection of spaces denoted by $(\{X_\alpha\}_{\alpha\in A})$. If a collection of sets$(\{B_\alpha\}_{\alpha\in A})$ serves as topology basis of each (X_α) where$(\alpha \in A)$, then the collection $\left(\left\{\prod_{\alpha\in A} B_\alpha \mid B_\alpha = X_\alpha \text{ for all but finitely many } \alpha\right\}\right)$ can be considered as a basis for the product topology on $\left(\prod_{\alpha\in A} X_\alpha\right)$.

If we have a nonempty open subset $(G) of \left(\prod_{\alpha\in A} X_\alpha\right)$ and $(x \in G)$, then one can find open sets of a finite number $(U_\alpha \subseteq X_\alpha), (i=1,\ldots,n)$, such that $\left(x \in \prod_{\alpha\in A} U_\alpha \subseteq G\right)$. We can take $(B_\alpha = X_\alpha)$ for $(\alpha \neq \alpha_1, \ldots, \alpha_n)$, in the function $\left(p_\alpha : \prod_{\alpha\in A} X_\alpha \to X_\alpha\right)$. This gives us $\left(x \in \prod_{\alpha\in A} B_\alpha = \bigcap_{i=1}^{n} p_\alpha^{-1}(B_\alpha) \subseteq G\right)$, and thus the proposition is proven.

Proposition 2.2.9: Consider a collection of spaces denoted by $(\{X_\alpha\}_{\alpha\in A})$ and let Y be a subset of the Cartesian product of these spaces, $\left(\prod_{\alpha\in A} X_\alpha\right)$. The topology of the product space $\left(\prod_{\alpha\in A} X_\alpha\right)$ induces a relative topology, which is similar to product topology for (Y), where each$(Y_\alpha = Y \cap X_\alpha)$ has the subspace topology.

Theorem 2.2.10 : Let be a family of spaces $(\{X_\alpha\}_{\alpha\in A})$. The continuity of a function (f) from a space (Y) into the product space $(\prod_{\alpha\in A} X_\alpha)$ depends on the continuity of the composition $(p_\alpha \circ f)$ for each $(\alpha \in A)$. Furthermore, the product topology is characterized by this characteristic.

Proof: Each $(p_\alpha \circ f)$, which is the product of two continuous maps, is continuous if (f) is continuous. On the other hand, assume that for any (alpha in A), $(p_\alpha \circ f)$ is continuous.

Since $(p_\alpha^{-1}(U_\alpha)),(U_\alpha \subseteq X_\alpha)$ open, they form a subbasis for the continuous space $\left(\prod_{\alpha\in A} X_\alpha\right),(f)$.

In order to observe the final statement, we define the product space $\left(\prod_{\alpha\in A} X_\alpha\right)$ as (X), and we denote the set $\left(\prod_{\alpha\in A} X_\alpha\right)$ with a topology as $(\ )$. In this topology, a function (f) from a space (Y) into (X_r) is considered continuous if the composition $(p_\alpha \circ f)$ from (Y) to(x)is continuous for every(α). Each projection (p_α) of (X_r) onto $(X_\alpha)$is continuous since the (X_r) identity map is continuous. Hence, functions $(X \xrightarrow{i} X_r \xrightarrow{p_\alpha} X_\alpha)$ and $(X_r \xrightarrow{i^{-1}} X \xrightarrow{p_\alpha} X_\alpha)$, where $(i : X \to X_r)$ represents the identity function, are continuous for any (α). Therefore, topologies of (X) and $(X_r)$are identical (Dugré et al., 2016).

The function $\left(f : Y \to \prod_{\alpha\in A} X_\alpha\right)$, can be described as having the map $(p_\alpha \circ f : Y \to X_\alpha)$ as its $(\alpha)th$ coordinate function (f). The previously mentioned theorem is useful for creating Cartesian products from continuous functions. Indeed, if continuous mapping $(f_\alpha : Y \to X_\alpha)$ is provided for every $(\alpha \in A)$, then the function from (Y) to $\left(\prod_{\alpha\in A} X_\alpha\right)$ is continuous. Another application of this concept is that if $(f : X \to Y)$ is a family of continuous functions, then the product function $(f_\alpha : X \to Y_\alpha)$, which maps $(x_\alpha))$into $((f_\alpha(x_\alpha))$, is also continuous.

In relation to the products of metric spaces, we investigate whether there is a metric that generates the product topology on the collection of metric spaces $(\{X_\alpha\}_{\alpha\in A})$, and their Cartesian product $\left(\prod_{\alpha\in A} X_\alpha\right)$. Yes, the answer is positive for finite products. If $(X_i, d_i), 1 \le i \le n$, is a finite collection of metric spaces, then the function (ρ) defined as

$$\rho((x_i),(y_i)) = \max\{d_i(x_i, y_i) \mid 1 \le i \le n\}$$

An open ball $(B((x_i),r))$ in the metric $(\prod_{i=1}^{n} X_i)$ can be observed as the product of open balls $B((x_i),r)$ in $((x_i))$. Additionally, an open balls $product(B(x_i,r_i),1 \le i \le n)$, includes open ball $B((x_i),r)$, where $(r = \min\{r_i \mid 1 \le i \le n\})$.

If coordinate spaces diameters are constrained to a particular value, the previously stated definition of metric can be expanded to encompass an infinite product of metric spaces. In metric spaces context, if we have a family of metric spaces $((X_\alpha, d_\alpha), \alpha \in A)$, where each space has a diameter at most (k) (a constant), we can define a metric on the set $\left(\prod_{\alpha \in A} X_\alpha\right)$ using the function $(\rho((x_\alpha),(y_\alpha)) = \sup\{d_\alpha(x_\alpha, y_\alpha) \mid \alpha \in A\})$.

It can be shown that topology (T_ρ) produced by metric (ρ), is more refined than the product topology, much like finite products. On other hand, it's important to note that the converse may not always hold. When considering countable products, it is interesting to note that the induced topology (T_ρ) aligns with the product topology, with a small adjustment in the definition of (ρ),.

Theorem 2.2.11 Consider a countable collection of metric spaces denoted by $(\{X_n\}_{n \in N})$. $(\prod_{n \in N} X_n)$ product topology is generated by a metric.

Proof: For every (n), select a positive real integer (λ_n) and $(\lambda_n \to 0)$. According to Corollary 1.4.9, one can infer that each (X_n) is associated with a metric (d_n) such that the $that(\operatorname{diam}(X_n) \le \lambda_n)$. The function $\left(\rho((x_n),(y_n)) = \sup_{n \in N} d_n(x_n, y_n)\right)$ for (n) in the set of natural numbers. Let $(\rho(x, y))$ be a nonnegative real number. It can be easily confirmed that (ñ) is a metric on $(\prod_{n \in N} X_n)$. We demonstrate that it produces the product topology (Lee net al., 2001).

A fundamental open set (U) in the product topology can be expressed as $(U = \prod_{n=1}^{k} G_n)$, where (G_n) is an open set in (X_n) for $(n=1,\ldots,k)$ and $(G_n = X_n)$ for all other indices (n).

Let $(x \in U)$. Select open balls $(B(x_n, r_n) \subseteq G_n)$ for $(n=1,2,\ldots,k)$, and define $(r = \min(r_1,\ldots,r_k))$. If $(r>0)$ and $(\rho(x,y)<r)$, then $(y_n \in B(x_n, r_n))$ for each $(n=1,\ldots,k)$. Consequently, it can be deduced that $(B(x,r) \subseteq U)$. Proposition 1.4.6 states that product topology is less complex as compared to metric topology produced by (ρ).

On the other hand, consider $(B(x,r))$ which is an open ball in the metric space (X_n). Given that the sequence $(\mathrm{diam}(X_n) \to 0)$, an integral is there (n_0) such that $(\mathrm{diam}(X_n) < r/2)$. Consider the open ball G_n defined as $(B(x, r/2) \setminus$ for $(n \leq n_0)$, and let $(G_n = X_n)\, forall (n > n_0)$. Let $(U = \prod_{n \in N} G_n)$. In the product topology, (U) is a fundamental open neighborhood of (x) and it is contained within the open ball $(U \subseteq B(x,r))$. As a result, the two topologies are the same, with the metric topology being less refined than the product topology.

Definition 2.2.12: In order for a topological space (X) to be considered metrizable, it must have a metric that generates its topology. Theorem 2.2.11 can be restated using this terminology in the following manner: If a metrizable space is a countable product of other metrizable spaces, then it is metrizable3. This conclusion is crucial for demonstrating a significant theorem in Chapter 8, which establishes the necessary conditions for a space to be metrizable.

If $\{X_\alpha\}_{\alpha \in A}$ is a set of countably infinite sets where each $(X_\alpha = X)$, we often denote (X^ω) as (X^A), as only the size of (A) matters in calculating (X^A). Consequently, the product space (X^ω) is metrizable for any metric space (X);. Specifically, (R^ω) and (l^ω) are examples of metrizable spaces. It is crucial to note that the metrizability of the product of an unlimited number of metric spaces, each containing more than one point, is not possible. The reason for this is that the product lacks a countable local basis at any given location, as stated in Theorem 7.1.7. On other hand, open balls with radii $(1/n),(n=1,2,\ldots)$ around particular point in metric space serves as a countable local basis (Hauser et al., 2007).

ACTIVITY 2.1.

Explore the concept of infinite product topologies and their applications, such as in infinite-dimensional spaces and functional analysis.

SUMMARY

Chapter 2 explores the fundamental concepts of continuity and product topology in the field of topology. The chapter begins with a thorough examination of continuous functions, starting with their formal definition and progressing through various equivalent conditions that characterize continuity. It also explains the role of bases and subbases in establishing continuity, providing numerous examples to clarify these concepts.

The discussion extends to the continuity of identity and composition maps, emphasizing their significance in different topological contexts. The concept of the inclusion map and the relative topology is introduced, highlighting their importance in understanding localized continuity. The chapter also presents the Gluing Lemma, a crucial result that allows for the construction of continuous functions from locally defined continuous maps

Further, the chapter differentiates between closed maps and closure operations, examining their interaction with continuity. It provides a detailed analysis of open maps and their essential role in the structure of topological spaces. The concept of topological embeddings is discussed, demonstrating their importance in embedding one topological space into another.

In the second section, the focus shifts to product spaces and product topology. The chapter defines product topology and explores its foundational properties, including its basis. It addresses the geometric interpretations and higher-dimensional aspects of product topology, emphasizing the significance of associativity and commutativity properties. The chapter also highlights the importance of product topology in various mathematical contexts, particularly in the study of infinite products and their associated topologies.

MULTIPLE CHOICE QUESTIONS

1. **In the context of topological spaces, a function f: $X \to Y$ is continuous if:**
   a. The inverse image of every open set in Y is open in X
   b. The image of every open set in X is open in Y.
   c. The inverse image of every closed set in Y is closed in X.
   d. The image of every closed set in X is closed in Y
2. **The Gluing Lemma is used to:**
   a. Combine closed sets to form open sets.
   b. Construct continuous functions from locally defined continuous functions.
   c. Identify continuous functions that are also open maps.
   d. Determine the closure of a set in a given topology.
3. **Which of the following properties is NOT typically associated with the product topology?**
   a. Basis elements are products of basis elements from the component topologies.
   b. Product topology is associative and commutative.

c. The product of compact spaces is always compact.

d. Infinite products always result in a Hausdorff space.

4. What is the basis of the product topology for a Cartesian product of spaces?

a. Closed sets

b. Open sets

c. Compact sets

d. Bounded sets

5. In the product topology, what property ensures that projections are continuous?

a. Sequential compactness

b. Separability

c. Tychonoff theorem

d. Limit point compactness

6. Which property characterizes the continuity of the inclusion map in relative topology?

a. Injectivity

b. Surjectivity

c. Openness

d. Closedness

REVIEW QUESTIONS

1. Define a continuous function between two topological spaces. What are the equivalent conditions that characterize continuity?
2. How do bases and subbases contribute to determining the continuity of a function?
3. What is the Gluing Lemma, and how does it apply to constructing continuous functions?
4. Differentiate between closed maps and open maps in the context of topological spaces.
5. How is product topology defined on a Cartesian product of topological spaces, and what are its fundamental properties?
6. What is a topological embedding, and what role does it play in preserving topological properties?

Answers to Multiple Questions

1\. (a) 2. (b) 3. (d) 4. (b) 5. (c)

6\. (c)

REFERENCES

1. Ajmal, N., & Sarma, R. D. (1992). Various fuzzy topological notions and fuzzy almost continuity. *Fuzzy Sets and Systems*, *50*(2), 209–223.
2. Alimi, I. A., Patel, R. K., Aboderin, O., Abdalla, A. M., Gbadamosi, R. A., Muga, N. J.,... & Teixeira, A. L. (2022). Network-on-Chip Topologies: Potentials, Technical Challenges, Recent Advances and Research Direction. *Network-on-Chip: Architecture, Optimization, and Design Explorations*, 39, 652–655.
3. Aste, T., Gramatica, R., & Di Matteo, T. (2012). Exploring complex networks via topological embedding on surfaces. *Physical Review E—Statistical, Nonlinear, and Soft Matter Physics*, *86*(3), 036109.
4. Baker, C. W. (1986). Characterizations of some near-continuous functions and near-open functions. *International Journal of Mathematics and Mathematical Sciences*, *9*(4), 715–720.
5. Baraglia, D., & Konno, H. (2020). A gluing formula for families Seiberg–Witten invariants. *Geometry & Topology*, *24*(3), 1381–1456.
6. Bhardwaj, M., Tyagi, B. K., & Singh, S. (2020). Relativization of star-C-Hurewicz property in topological spaces. *Bulletin of the Iranian Mathematical Society*, *46*(4), 1101–1115.
7. Bolsinov, A., & Matveev, V. (2011). Splitting and gluing lemmas for geodesically equivalent pseudo-Riemannian metrics. *Transactions of the American Mathematical Society*, *363*(8), 4081–4107.
8. Bonner, S., Brennan, J., Kureshi, I., Theodoropoulos, G., McGough, A. S., & Obara, B. (2017, December). Evaluating the quality of graph embeddings via topological feature reconstruction. In *2017 IEEE International Conference on Big Data (Big Data)*. IEEE. (Vol. 2, pp. 2691–2700).
9. Bonsangue, M. M., van Breugel, F., & Rutten, J. J. (1998). Generalized metric spaces: completion, topology, and powerdomains via the Yoneda embedding. *Theoretical Computer Science*, *193*(1–2), 1–51.
10. Brown, R. (1964). Function spaces and product topologies. *The Quarterly Journal of Mathematics*, *15*(1), 238–250.
11. Bunke, U. (1995). On the gluing problem for the η-invariant. *Journal of Differential Geometry*, *41*(2), 397–448.
12. Caspard, N., & Monjardet, B. (2003). The lattices of closure systems, closure operators, and implicational systems on a finite set: a survey. *Discrete Applied Mathematics*, *127*(2), 241–269.
13. Chittenden, E. W. (1929). On general topology and the relation of the properties of the class of all continuous functions to the properties of space. *Transactions of the American Mathematical Society*, *31*(2), 290–321.
14. Ciesielski, K., & Rosłanowski, A. (2000). Two examples concerning almost continuous functions. *Topology and its Applications*, *103*(2), 187–202.
15. Cincio, L., & Vidal, G. (2013). Characterizing topological order by studying the

ground states on an infinite cylinder. *Physical review letters*, *110*(6), 067208.

16. Comfort, W. W., & Grant, D. L. (1981). Cardinal invariants, pseudocompactness and minimality: some recent advances in the topological theory of topological groups. In *Topology Proc* (Vol. 6, pp. 227–265).
17. Comfort, W. W., & Negrepontis, S. (1972). Continuous functions on products with strong topologies. *General Topology and its Relations to Modern Analysis and Algebra*, 89–92.
18. Davoren, J. M. (1999). Topologies, continuity and bisimulations. *RAIRO-Theoretical Informatics and Applications*, *33*(4–5), 357–381.
19. Dedushenko, M. (2020). Gluing. Part I. Integrals and symmetries. *Journal of High Energy Physics*, *2020*(4), 1–48.
20. Di Maio, G., Holý, D., & McCoy, R. A. (1998). Topologies on the space of continuous functions. *Topology and its Applications*, *86*(2), 105–122.
21. Dikranjan, D., & Giuli, E. (1987). Closure operators I. *Topology and its Applications*, *27*(2), 129–143.
22. Dimofte, T., & Garoufalidis, S. (2013). The quantum content of the gluing equations. *Geometry & Topology*, *17*(3), 1253–1315.
23. Drewnowski, L. (2007). Resolutions of topological linear spaces and continuity of linear maps. *Journal of mathematical analysis and applications*, *335*(2), 1177–1194.
24. Dubrovin, B. A., Fomenko, A. T., & Novikov, S. P. (2012). *Modern geometry—methods and applications: Part II: The geometry and topology of manifolds.* Springer Science & Business Media. (Vol. 104, pp. 25–30).
25. Dugré, A., Vadean, A., & Chaussée, J. (2016). Challenges of using topology optimization for the design of pressurized stiffened panels. *Structural and Multidisciplinary Optimization*, 12. *53*, 303–320.
26. Elliott, L., Jonušas, J., Mesyan, Z., Mitchell, J., Morayne, M., & Péresse, Y. (2023). Automatic continuity, unique Polish topologies, and Zariski topologies on monoids and clones. *Transactions of the American Mathematical Society*, *376*(11), 8023–8093.
27. Fabrizi, E., & Saffiotti, A. (2000, April). Extracting topology-based maps from gridmaps. In *Proceedings 2000 ICRA. Millennium Conference. IEEE International Conference on Robotics and Automation. Symposia Proceedings (Cat. No. 00CH37065).* IEEE. (Vol. 3, pp. 2972–2978).
28. Forstneric, F., & Rosay, J. P. (1987). Localization of the Kobayashi metric and the boundary continuity of proper holomorphic mappings. *Mathematische Annalen*, *279*, 239–252.
29. Gallagher, K., & Engelke Infante, N. (2023). Generic examples in undergraduate topology: a case study. *International Journal of Mathematical Education in Science and Technology*, *54*(9), 1833–1852.
30. Ganster, M., & Reilly, I. L. (1989). Locally closed sets and LC-continuous functions. *International Journal of Mathematics and Mathematical Sciences*, *12*(3), 417–424.
31. Garoufalidis, S., Goerner, M., & Zickert, C. (2015). Gluing equations for PGL (n, C)–representations of 3–manifolds. *Algebraic & Geometric Topology*, *15*(1), 565–622.

32. Grabowski, A. (1999). Properties of the product of compact topological spaces. *Formalized Mathematics*, *8*(1), 55–59.

33. Habib, S. J. (2008). Redesigning network topology with technology considerations. *International Journal of Network Management*, *18*(1), 1–13.

34. Hadjigeorgiou, R. I. (1999). Gluing spectra and boundaries of topological algebras together. *Journal of Mathematical Sciences*, *95*(6), 2626–2637.

35. Hauser, H., Laramee, R. S., & Doleisch, H. (2007). Topology-based versus Feature-based Flow Analysis–Challenges and an Application. In *Topology-based Methods in Visualization*. Berlin, Heidelberg: Springer Berlin Heidelberg. (Vol. 5, pp. 79–90).

36. Heaton, P., Miles, S., & Duhan, S. (2012). Dynamic mapping of stakeholders for dealing with plant closure complexity. *Social Business*, *2*(2), 95–119.

37. Henriksen, M., & Woods, R. G. (1999). Separate versus joint continuity: A tale of four topologies. *Topology and its Applications*, *97*(1–2), 175–205.

38. Herbst, M., & Lazaroiu, C. I. (2005). Localization and traces in open-closed topological Landau-Ginzburg models. *Journal of High Energy Physics*, *2005*(05), 044.

39. Huang, J. (2011). A gluing construction for polynomial invariants. *Journal of Algebra*, *328*(1), 432–442.

40. Itzkowitz, G. L. (2020). Uniformities and uniform continuity on topological groups. In *General Topology and Applications*. CRC Press. (Vol. 3, pp. 155–178).

41. Jaffe, A., Lee, J. R., & Moharrami, M. (2011). On the optimality of gluing over scales. *Discrete & Computational Geometry*, *46*(2), 270–282.

42. Kamae, T., Luo, J., & Tan, B. (2015). A gluing lemma for iterated function systems. *Fractals*, *23*(02), 1550019.

43. Kowalczyk, S., & Turowska, M. (2021). On continuity in generalized topology. *Topology and its Applications*, *297*, 107702.

44. Larsson, R., & Runesson, K. (1996). Element-embedded localization band based on regularized displacement discontinuity. *Journal of Engineering Mechanics*, *122*(5), 402–411.

45. Lee, F. C., Barbosa, P., Xu, P., Zhang, J., Yang, B., & Canales, F. (2001). Topologies and design considerations for distributed power system applications. *Proceedings of the IEEE*, *89*(6), 939–950.

46. Liu, X., Zhuang, C., Murata, T., Kim, K. S., & Kertkeidkachorn, N. (2019). How much topological structure is preserved by graph embeddings? *Computer Science and Information Systems*, *16*(2), 597–614.

47. Long, P. E., & Herrington, L. L. (1982). The T θ-topology and faintly continuous functions. *Kyungpook Mathematical Journal*, *22*(1), 7–14.

48. Martynyuk, V. V. (1971). Some applications of the closure operation in graph analysis. *USSR Computational Mathematics and Mathematical Physics*, *11*(6), 320–324.

49. Mayer, W. (1945). On products in topology. *Annals of Mathematics*, *46*(1), 29–57.

50. Møller, J. M. (2007). General topology. *Matematisk Institut, Universitetsparken,* 5, 1-16.

51. Naimpally, S. A. (1966). Graph topology for function spaces. *Transactions of the American Mathematical Society*, *123*(1), 267–272.

52. Niu, Z. (2022). Comparison radius and mean topological dimension: Rokhlin property, comparison of open sets, and subhomogeneous C*-algebras. *Journal d'Analyse Mathématique*, *146*(2), 595–672.

53. Núñez, J., González, A., & Ramos, E. (2022). Topological data analysis of Lagrangian orbits in natural convection flows confined in a cylinder. *Physical Review Fluids*, *7*(12), 123501.

54. Preuss, G. (2012). *Theory of topological structures: an approach to categorical topology.* Springer Science & Business Media. (Vol. 39, pp. 5–10).

55. Radó, T., & Reichelderfer, P. V. (2012). *Continuous transformations in analysis: with an introduction to algebraic topology.* Springer Science & Business Media. (Vol. 75, pp. 65–69).

56. Rashevsky, N. (1954). Topology and life: in search of general mathematical principles in biology and sociology. *The bulletin of mathematical biophysics*, *16*, 317–348.

57. Rounds, W. C. (1984, July). Applications of topology to semantics of communicating processes. In *International Conference on Concurrency.* Berlin, Heidelberg: Springer Berlin Heidelberg. (Vol. 9, pp. 360–372).

58. Sayre, R., Frye, C., Karagulle, D., Krauer, J., Breyer, S., Aniello, P.,... & Cress, J. (2018). A new high-resolution map of world mountains and an online tool for visualizing and comparing characterizations of global mountain distributions. *Mountain Research and Development*, *38*(3), 240–249.

59. Schmidt, G., & Berghammer, R. (2011). Contact, closure, topology, and the linking of row and column types of relations. *The Journal of Logic and Algebraic Programming*, *80*(6), 339–361.

60. Sénizergues, D. (2019). Random gluing of metric spaces. *The Annals of Probability*, *47*(6), 3812–3865.

61. Sgro, J. (1976). Completeness theorems for continuous functions and product topologies. *Israel Journal of Mathematics*, *25*, 249–272.

62. Smyth, M. B. (1995). Semi-metrics, closure spaces and digital topology. *Theoretical computer science*, *151*(1), 257–276.

63. Torkestani, J. A. (2013). An energy-efficient topology construction algorithm for wireless sensor networks. *Computer Networks*, *57*(7), 1714–1725.

64. Westerveld, L., & Knowles, A. K. (2021). Loosening the grid: topology as the basis for a more inclusive GIS. *International Journal of Geographical Information Science*, *35*(10), 2108–2127.

65. Yuan, G. X. Z. (1999). The characterization of generalized metric KKM mappings with open values in hyperconvex metric spaces and some applications. *Journal of mathematical analysis and applications*, *235*(1), 315–325.

CHAPTER

3

Compactness and Connectedness

LEARNING OBJECTIVES

After studying this chapter, you will be able to:

1. Understand the definition and basic properties of compact spaces.
2. Explore the Heine-Borel Theorem and its applications.
3. Analyze compactness in product spaces and its implications.
4. Learn about sequential compactness and its significance in metric spaces.
5. Characterize compactness in metric spaces and its relation to continuous functions.
6. Comprehend the definition and basic properties of connected and path-connected spaces.
7. Identify and differentiate between components and path components.
8. Apply concepts of compactness and connectedness to various advanced topics in analysis and topology.

KEY TERMS FROM THIS CHAPTER

Compact Space
Product Space
Limit Point Compactness
Components
Topological Groups
Heine-Borel Theorem
Sequential Compactness
Path-Connected Space
Path Components

UNIT INTRODUCTION

Compactness and connectedness are fundamental concepts in topology that provide deep insights into the structure and behavior of topological spaces. Compactness refers to a space in which every open cover has a finite subcover, indicating that the space can be efficiently covered using a finite number of open sets. This property is crucial in various areas of mathematics, as it often ensures the existence of limits, maxima, and minima within space. For example, in real analysis, the Heine-Borel Theorem characterizes compact subsets of Euclidean space, establishing a foundational connection between topology and analysis. Compactness is also vital in functional analysis, particularly in the study of function spaces, where it supports important theorems such as the Arzelà-Ascoli Theorem and the Banach-Alaoglu Theorem (Al-shami et al., 2023).

Connectedness describes a space that cannot be divided into two disjoint non-empty open subsets, implying that the space is "in one piece." This concept is essential for understanding the continuity and behavior of functions across space. Connectedness plays a crucial role in various mathematical theorems, such as the Intermediate Value Theorem in real analysis, which depends on the connectedness of intervals. In algebraic topology, connectedness is central to the study of fundamental groups and homotopy, providing a framework for classifying and understanding topological spaces based on their path-connected components. These properties are not only theoretical foundations but also have practical applications in fields such as data analysis, quantum physics, and network theory, underscoring their broad relevance in both pure and applied mathematics (Azarpanah et al., 2015).

3.1. COMPACT SPACES

Following is the explanation of compact spaces.

3.1.1. Definition of Compact Spaces

Following is the definition and example of compact spaces:

3.1.1.1. Definition of Compact Spaces

A topological space given by (X, τ) is considered compact if each of the open cover of (X) can be reduced to a finite subcover (Figure 3.1).

In formal terms, if $\{U_\alpha\}_{\alpha \in A}$ is an open cover of (X), it results that there occurs a finite subset $\{U_{\alpha_1}, U_{\alpha_2}, \ldots, U_{\alpha}\}$ such that $\left(X = \bigcup_{i=1}^{n} U_{\alpha_i}\right)$.

Mathematically, this can be expressed as (Geba, 1964):

For X

$\bigcup_{i=1}^{n} U_{\alpha_i}$.

Remember

Remember the definition and key properties of compact spaces, especially the Heine-Borel theorem which characterizes compactness in Euclidean spaces.

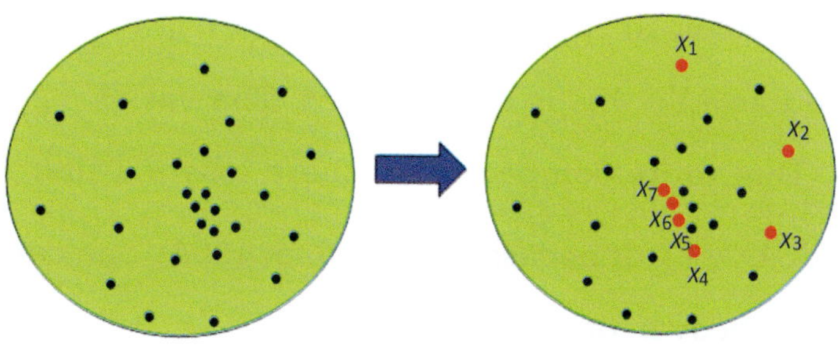

Figure 3.1. Example of compact spaces (Source: Math Strek, Creative Commons License)

3.1.1.2. Examples of Compact Spaces

1. **Closed Interval in $(\mathbb{R})$**

It is compact to have a closed interval $(a,\subset R)$. There is a compact set represented by the closed interval $(a,\subset R)$. According to the Heine-Borel Theorem, a subset of (Rn) is compact if and only if it meets the requirements of being both closed and bounded. The set $(a,)$ can be categorized as compact since it meets the requirements for being both closed and bounded (Bekkali et al., 1992).

2. Finite Sets

Every finite set, regardless of its topology, is compact. For each finite set $\left(X=\{x_1,x_2,\ldots,x_n\}\right)$ and any topology on (X), every open cover must include at least one open set containing each element (Vaughan, 1984). Therefore, the finite union of these open sets will cover (X).

3. Compact Subsets of $(\boldsymbol{R}^n)$

Every subset of (R^n) that is both closed and limited is also compact. The closed ball $\left(B=\{x \in R^n \mid |x| \leq r\}\right)$ is compact because it satisfies the properties of being both closed and bounded (Tanaka, 1986).

3.1.2. Basic Properties of Compact Spaces

Following are the properties of compact spaces.

3.1.2.1. Closed Subsets of Compact Spaces

If (X,τ) is a compact space and $(Y \subseteq X)$ is a closed subset, then (Y) is also compact. To illustrate this, analyze an open cover $\{V_\beta\}_{\beta \in B}$ of (Y). Since (Y) is a closed set, the complement of (Y) in (X) denoted as $(X \setminus Y)$, is an open set. The set $\{V_\beta\}_{\beta \in B} \cup \{X \setminus Y\}$ serves as an open cover for (X). Due to its compactness, there exists a finite subcover composed of the sets $\{V_{\beta_1}, V_{\beta_2}, \ldots, V_{\beta_m}, X \setminus Y\}$. If necessary (Treybig, 1964), one can eliminate the set $(X \setminus Y)$ to create a finite subcover of (Y).

3.1.2.2. Continuous Image of Compact Spaces

Given any open cover $\{W_\gamma\}_{\gamma \in \Gamma})$ of $(f(X)$, the preimages $\{f^{-1}(W_\gamma)\}_{\gamma \in \Gamma}$ form an open cover of (X). By compactness, there is a finite subset that covers $\{f^{-1}(W_{\gamma_1}), f^{-1}(W_{\gamma_2}), \ldots, f^{-1}(W_{\gamma_n})\}$ (Kohli & Das, 2006). The corresponding finite collection $\{W_{\gamma_1}, W_{\gamma_2}, \ldots, W_{\gamma_n}\}) covers (f(X)$.

3.1.3. Heine-Borel Theorem

This theorem results in the domains of real analysis and topology.

Statement:

A subset $\left(K \subseteq R^{n}\right)$ is considered compact if and only if it satisfies the conditions of being closed and bounded (So, 1991).

Proof:

1. Necessity:

i. Suppose $\left(K \subseteq R^{n}\right)$ is compact.

ii. Bounded: Assume (K) is not bounded. Then, for each integer (n), consider the open cover $(\{B_n\})where(B_n = \{x \in R^n \mid |x| < n\})$. Since (K) is unbounded, no finite collection of (B_n) can cover (K), contradicting the compactness of (K). Thus, (K) must be bounded.

iii. Closed: Suppose (K) is not closed. Then there exists a limit point $\left(x \in \underline{K} \setminus K\right)$. Consider the open cover $\{B_\epsilon(x) \mid \epsilon > 0\}$ of (K) where $(B_\epsilon(x))$ denotes the open ball centered at (x) with radius (ϵ) Since $\left(x \notin K\right)$, no finite subcover can cover (K), contradicting the compactness. Thus, (K) must be closed (Moore, 1919).

2. Sufficiency:

i. Suppose $\left(K \subseteq R^{n}\right)$ is closed and bounded. It is needed to show $(\ )$ is compact.

ii. Bounded: Since (K) is bounded, it fits within some closed ball $\left(B \subseteq R^{n}\right)$.

iii. Closed: Since (K) is closed, it contains all its limit points.

iv. Compactness: Consider an open cover $(\{U_\alpha\}_{\alpha \in A})of(K)$. Since $\left(K \subseteq B\right)$, and (B) are compact in $\left(R^{n}\right)$, there exists a finite subcover of (B). This subcover also covers (K), proving (K) is compact (Barone & Novikoff, 1978).

3.1.4. Applications of the Heine-Borel Theorem

3.1.4.1. Extreme Value Theorem

The Heine-Borel Theorem plays a vital role in establishing the Extreme Value Theorem, which asserts that a continuous function $(f: K \rightarrow R)$ defined on a compact set (K) achieves its highest and lowest values. Given that (K) is a compact set, it follows that, $(f(K))$ is similarly compact (Fitch, 1950). Consequently, $(f(K))$ is closed and bounded, ensuring the existence of both a maximum value $(maxf)$ and a minimum value (f).

3.1.4.2. Uniform Continuity

The theorem establishes that any continuous function defined on a compact set possesses the attribute of uniform continuity. Given a continuous function $(f: K \rightarrow R)$ where (K) is a compact set, for any $(\dot{o}>0)$, there exists $(\delta>0)$ such that for all $(x, y \in K)$, $(|x-y|<\delta)$ implies $(|f(x)-f(y)|<\dot{o})$.

3.1.4.3. Compact Operators

In functional analysis, the Heine-Borel Theorem aids in defining compact operators. An operator $(T: X \rightarrow Y)$ (where (X) and (Y) are normed vector spaces) is compact if it translates sets that have a finite size to sets that are compact in comparison. This property is crucial in the study of integral equations and spectral theory (Sanders, 2021).

3.1.4.4. Finite-Dimensional Vector Spaces

The theorem's outcome is especially valuable in the realm of linear algebra and optimization, as the compactness of viable areas guarantees the presence of optimum solutions.

3.1.4.5. Covering Theorems

The theory serves as the basis for a number of covering theorems, including Lebesgue's Covering Lemma, which asserts that there is a finite open cover with arbitrarily tiny diameters for every open cover of a compact metric space.

The Heine-Borel Theorem thus plays a central role in both theoretical and applied mathematics, underpinning many key results and applications across various fields (Cederquist & Negri, 1995).

3.1.5. Compactness in Product Spaces

3.1.5.1. Tychonoff's Theorem

This theorem states that the product of any set of compact spaces is also compact.

Statement:

In a formal sense, if $(\{X_\alpha\}_{\alpha \in A})$ represents a collection of compact spaces, then their product $\left(\prod_{\alpha \in A} X_\alpha\right)$, which is given with the product topology, is also compact.

Proof:

Tychonoff's Theorem will be shown by using the idea of filters. An equivalent form of compactness is that every filter on a space (X) has a cluster point (Rettinger & Weihrauch, 2013).

Did you know?

Did you know that in product spaces, a product of compact spaces is compact under the product topology, which is a fundamental result in topology?

1. Filters and Ultrafilters:

A filter (F) on a set (X) is a non-empty collection of subsets of (X) that is closed under finite intersections and supersets (Lawrence, 1990).

An ultrafilter (U) on (X) is a maximal filter, meaning that for any subset $(A \subseteq X)$, either $(A \in U) or (X \setminus A \in U)$.

2. Compactness via Ultrafilters:

A space (X) is considered compact if any ultrafilter on (X) converges to at least one point.

3. Product Spaces and Projections:

Consider the product space $\left(\prod_{\alpha \in A} X_\alpha\right)$ equipped with the product topology.

For each $(\alpha \in A), let \left(\pi_\alpha : \prod_{\alpha \in A} X_\alpha \to X_\alpha\right)$ be the projection map (Comfort & Hager, 1971).

4. Constructing an Ultrafilter on the Product Space:

Given an ultrafilter (U) on $\left(\prod_{\alpha \in A} X_\alpha\right)$, it is needed to show (U) converges to a point in $\left(\prod_{\alpha \in A} X_\alpha\right)$.

5. Ultrafilters on Component Spaces:

For each $(\alpha \in A)$, define $\left(U_\alpha = \{\pi_\alpha(F) \mid F \in U\}\right)$.

(U_α) is an ultrafilter on = because the projection map π_α is continuous.

6. Compactness of (X_α):

Since each (X_α) is compact (Podczeck, 2008), the ultrafilter (U_α) converges to some point $(x_\alpha \in X_\alpha)$.

7. Defining the Limit Point:

Define the point $\left(x = (x_\alpha)_{\alpha \in A} \in \prod_{\alpha \in A} X_\alpha\right)$.

8. Convergence in the Product Space:

To demonstrate the convergence of (U) to

(x), it is necessary to confirm that for any open set (U) in $\left(\prod_{\alpha \in A} X_\alpha\right)$ that contains (x), there is $(U \in U)$.

9. Basic Open Sets in Product Topology:

A basic open set in the product topology is of the form $\left(\prod_{\alpha \in A} U_\alpha\right)$, where (U_α) is open in (X_α) and $(U_\alpha = X_\alpha)$ for all but finitely many (α).

10. Convergence to (x):

Since each (U_α) converges to (x_α), for any basic open set $\left(\prod_{\alpha \in A} U_\alpha\right)$ containing (x), the ultrafilter (U) must contain $\left(\prod_{\alpha \in A} U_\alpha\right)$.

Therefore, the ultrafilter (U) converges to (x), proving that the product space $\left(\prod_{\alpha \in A} X_\alpha\right)$ is compact (Pang & Shi, 2014).

3.1.5.2. Applications of Tychonoff's Theorem

1. Functional Analysis:

In the field of functional analysis, Tychonoff's Theorem is an important conclusion, particularly when it comes to the investigation of dual spaces and the Banach-Alaoglu Theorem. This is the case when the space is normed. In order to comprehend the Hahn-Banach Theorem and bounded linear functionals, it is necessary to become familiar with this theorem (Einsiedler & Ward, 2017).

2. Probability Theory:

The theorem is used in probability theory to establish the presence of product measures. In the case of a family of probability measures $(\{\mu_\alpha\}_{\alpha \in A})$ on compact spaces $(\{X_\alpha\}_{\alpha \in A})$, the coherence of multidimensional distributions is ensured by the existence of the product measure on the product space $(\prod_{\alpha \in} X_\alpha)$.

3. General Topology:

The theorem serves as the foundation for the concept of compactness in general topology, allowing for the extension of compactness criteria from finite products to products of any size. This generalization is essential for comprehending intricate topological structures and their characteristics in higher dimensions (Paterson & Welch, 2005).

4. Equicontinuity and Arzelà-Ascoli Theorem:

Tychonoff's Theorem is used to establish the Arzelà-Ascoli Theorem, which characterizes the compact subsets of the space of continuous functions. This statement claims that a group of continuous functions, which are uniformly bounded and equicontinuous, on a compact space is relatively compact if and only if it satisfies these conditions (Kumar, 2023).

5. Algebraic Topology:

In algebraic topology, the theorem is used in the study of compactly generated spaces, which are essential for homotopy theory and the development of various homological and cohomological theories. Compactly generated spaces allow for the extension of topological concepts to more abstract settings, facilitating the study of continuous mappings and their properties (Davis & Kirk, 2001).

Tychonoff's Theorem is thus a pivotal result, enabling a wide range of applications across different areas of mathematics, from pure topology to applied fields like functional analysis and probability theory (Massey, 2019).

3.1.6. Sequential Compactness/

Sequential compactness is a concept in topology that connects compactness with sequences. Sequential compactness and compactness are comparable in metric spaces, although this equivalence does not hold in all topological spaces (Juhász & Szentmiklóssy, 1993).

When $((X,\tau))$ is to be sequentially compact, it is because every sequence in (X) has a subsequence that converges to a limit inside (X).

According to the formal definition, a sequence (X) is considered sequentially compact if, for each sequence $(\{x_n\}_{n=1}^{\infty} \subseteq X)$, there exists a subsequence and a point $(x \in X)$ such that $(x_{n_k} \to x)$ as $(k \to \infty)$.

3.1.6.1. Relation to Compactness

Compactness and sequential compactness are two concepts that are comparable in metric spaces. As a result, a metric space $((X,d))$ may be deemed compact if and only if it demonstrates linear compactness.

Proof:

1. Compact implies Sequentially Compact:

For (X) consider a sequence $(\{x_n\}_{n=1}^{\infty})$ in the set (X). If there is a point ($\in X$) such that infinitely many terms of the sequence are in every neighborhood of (x), then (x) is a limit point of the sequence (Keremedis, 2016). It may be derived from a subsequence \ $(\{x_{n_k}\}_{k=1}^{\infty})$ that converges to (x).

2. Sequentially Compact implies Compact:

Assume (X) is sequentially compact. Let $(\{U_\alpha\}_{\alpha \in A})$ be an open cover of (X). To show (X) is compact, assume for contradiction that there is no finite subcover. Construct a sequence $(\{x_n\}_{n=1}^{\infty})$ where each (x_n) is chosen such that it is not covered by any finite subcollection of $(\{U_\alpha\}_{\alpha \in A})$. By sequential compactness, there exists a convergent subsequence $(\{x_{n_k}\}_{k=1}^{\infty})$ converging to some $(x \in X)$. *Since*$(\{U_\alpha\}_{\alpha \in A})$ is an open cover, (x) must lie in some (U_β), and all terms of the subsequence sufficiently close to (x) will be in (U_β), contradicting the construction of the sequence. Hence, a finite subcover must exist, proving compactness (Gänssler, 1971).

3.1.6.2. Sequentially Compact Spaces

Following are some examples of this:

1. Closed Interval in (R):

The closed interval $([a,b])$ contained in the set of real numbers $([a,b] \subset R)$ is sequentially compact. Every sequence inside the interval

$([a,b])$. has a subsequence that is bounded. According to the Bolzano-Weierstrass Theorem, this bounded subsequence also includes a subsequence that converges to a limit within the interval $([a,b])$.

2. Compact Subsets of (R^n)*:*

Any closed, bounded subset of (R^n) is sequentially compact (Haydon, 1977).

Properties of Sequentially Compact Spaces:

i. Limit Points: In a sequentially compact space, every infinite subset has a limit point. This follows because any infinite subset can be viewed as an infinite sequence, which must have a convergent subsequence.

ii. Boundedness: In metric spaces, sequential compactness implies boundedness. If $((X,d))$ is sequentially compact, then any sequence must have a convergent subsequence, implying the space cannot be unbounded.

iii. Closed Sets: Sequentially compact subsets of Hausdorff spaces are closed. In a Hausdorff space, the limits of sequences are unique, ensuring that the limit of any convergent subsequence lies within the set, thus making it closed (Joseph et al., 1999).

Sequential Compactness/ in Function Spaces:

i. Arzelà-Ascoli Theorem: The Arzelà-Ascoli Theorem offers a description of sequential compactness in spaces consisting of continuous functions.

ii. Application to Differential Equations: Sequential compactness is used in proving the presence of solutions to differential equations. For example, the Schauder Fixed Point Theorem leverages sequential compactness in infinite-dimensional spaces to establish the existence of fixed points (Nowak, 2011).

Implications in Analysis:

i. Uniform Convergence: In spaces of functions, sequential compactness ensures uniform convergence. If a sequence of functions converges pointwise and the space is sequentially compact, the convergence is uniform.

ii. Integration: Sequential compactness aids in the study of integrable functions. In (L^p) spaces, the compactness properties help in establishing results related to the convergence of sequences of functions under integration.

Sequential compactness thus provides a crucial link between sequences and topological properties, offering powerful tools for analysis and further study in both pure and applied mathematics (Wheeler, 1981).

3.2. COMPACTNESS IN METRIC SPACES

3.2.1. Characterization of Compactness

3.2.1.1. Bolzano-Weierstrass Theorem

The Bolzano-Weierstrass Theorem is a significant outcome in the field of real analysis and metric space theory, which provides a crucial definition of compactness in (R^n) (Figure 3.2).

Statement:

Every sequence that is contained inside a limited set (R^n) has a subsequence that converges.

Proof:

1. Boundedness:

Assume that $(\{x_n\})$ is a limited sequence in (R^n) (Colonna, 2013). Or $(M>0)$ such that $(|x_n| \leq M)$ for all (n).

2. Compact Subset:

Since $\{x_n\}$ is bounded, it lies within the closed ball of radius (M) centered at the origin, denoted as $(B_M(0) = \{x \in R^n \mid |x| \leq M\})$.

3. Closed and Bounded:

According to the Heine-Borel Theorem, the set $(B_M(0))$ is compact since it satisfies the conditions of being closed and bounded.

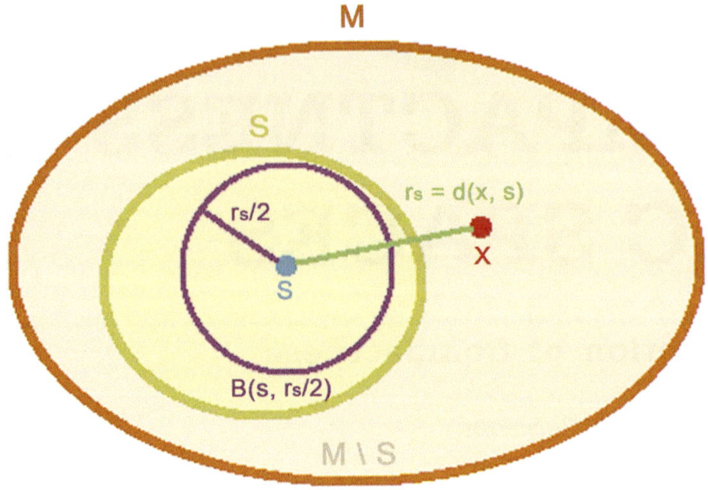

Figure 3.2. Closedness of Compact Sets in a Metric Space (Source: Math Online, Creative Commons License)

4. Limit Point:

Since $(B_M(0))$ is compact, every sequence in $(B_M(0))$ has a convergent subsequence. Therefore, $\{x_n\}$ has a convergent subsequence $(\{x_{n_k}\})$ converging to some point $x \in B_M(0)$.

This completes the Bolzano-Weierstrass Theorem's proof (Bugajewski & Gulgowski, 2020).

3.2.1.2. Sequential Characterization

Compactness in metric spaces may be defined in relation to sequences. This phenomenon is often known as sequential compactness.

Definition:

A metric space (X,d) is said to be progressively compact if every sequence in (X) has a subsequence that tends to converge to a limit inside (X).

Equivalence with Compactness:

Compactness and sequential compactness are identical in metric spaces. To clarify, a metric space (X,d) can only be considered compact if it exhibits sequential compactness (Ariza et al., 2000).

Proof:

1. *Compact implies Sequentially Compact:*

   i. Suppose (X,d) is compact. Let $\{x_n\}$ be any sequence in (X). It is needed to show that $(\{x_n\})$ has a convergent subsequence.

   ii. Consider the sequence $(\{x_n\})$ in (X). Every open cover associated with (X) has a finite subcover. In particular, one can cover (X) by open balls of radius $(\dot{o}=1)$.

   iii. Since (X) is compact, there exists a finite subcover of these open balls, say $(B_1, B_2, \ldots, B_k)$.

   iv. At least one of these balls must contain infinitely many terms of the sequence $(\{x_n\})$. Without loss of generality, assume (B_1) contains infinitely many terms.

   v. Repeat this process with $(\epsilon = \frac{1}{2}, \frac{1}{3}, \ldots)$ to extract a nested sequence of balls $(B_{1/2}, B_{1/3}, \ldots)$, each containing infinitely many terms.

   vi. By construction, the centers of these balls constitute a Cauchy sequence, converging to a point $(x \in X)$.

   vii. Hence, $(\{x_n\})$ has a subsequence converging to (x), proving sequential compactness (Font & Sanchis, 2015).

2. *Sequentially Compact implies Compact:*

   i. Suppose $((X,d))$ is sequentially compact. It is needed to show (X) is compact.

   ii. To avoid a contradiction, assume that the set (X) is not compact. Therefore, it may be said that there is an open cover $(\{U_\alpha\}_{\alpha \in A})$ of (X) that does not have any finite subcover.

   iii. Construct a sequence $(\{x_n\})$ in (X) such that no finite subcollection of $(\{U_\alpha\}_{\alpha \in A})$ covers the entire sequence.

   iv. By sequential compactness, $(\{x_n\})$ has a convergent subsequence $(\{x_{n_k}\})$ converging to some $(x \in X)$.

v. Since $(\{U_\alpha\}_{\alpha \in A})$ is an open cover, (x) must lie in some (U_β).

But infinitely many terms of $(\{x_{n_k}\})$ must also lie in (U_β), contradicting the construction of the sequence.

vi. Hence, (X) must be compact.

The purpose of these proofs is to prove the equivalence between compactness and sequential compactness in metric spaces (Mordukhovich & Wang, 2001).

Tip

Continuous functions preserve compactness; a continuous image of a compact space is compact, which is crucial in many applications, such as in analysis and functional analysis.

3.2.2. Compactness and Continuous Functions

3.2.2.1. Continuous Images of Compact Sets

One of the fundamental properties of compact sets in topology is their behavior under continuous mappings.

This relationship is crucial for understanding how compactness interacts with continuity.

Theorem:

A compact set has a continuous picture that is composed of a compact image.

Proof:

1. Setup:

    i. Let $((X, \tau_X))$ and $((Y, \tau_Y))$ be topological spaces.

    ii. Let $(f : X \to Y)$ as a continuous function.

    iii. Suppose $(K \subseteq X)$ is compact.

2. Goal:

    Show that $(f(K) \subseteq Y)$ is compact.

3. Covering by Open Sets:

    i. Let $(\{V_\alpha\}_{\alpha \in A})$ be an open cover of $(f(K))$.

    ii. Since (f) *is* continuous, $(f^{-1}(V_\alpha))$ is open in (X) for each $(\alpha \in A)$.

iii. The preimages $(\{f^{-1}(V_\alpha)\}_{\alpha\in A})$ form an open cover of (K).

4. Compactness of (K):

Since (K) is compact, there exists a finite subcover $\left(\left\{f^{-1}\left(V_{\alpha_1}\right), f^{-1}\left(V_{\alpha_2}\right), \ldots, f^{-1}\left(V_{\alpha_n}\right)\right\}\right)$.

5. Finite Subcover:

The corresponding sets $\left(\left\{V_{\alpha_1}, V_{\alpha_2}, \ldots, V_{\alpha_n}\right\}\right)$ form a finite cover of $(f(K))$.

Therefore, $(f(K))$ is covered by a finite subcollection of $(\{V_\alpha\}_{\alpha\in A})$, proving that $(f(K))$ is compact.

This theorem implies that compactness is preserved under continuous mappings, which is a powerful tool in various areas of mathematics (Bridges, 1978).

3.2.2.2. Extreme Value Theorem

In the field of real analysis, the Extreme Value Theorem is regarded as fundamental. The claim is that a continuous function formed on a compact set has a maximum value and a minimum value that it can reach. For any continuous function that does exist, this is the case (Cai & Daskalakis, 2015).

Theorem:

If $(f: X \rightarrow R)$ is a continuous function on a compact space (X), then (f) attains its maximum and minimum values on (X).

Proof:

1. Setup:

Let $(f: X \rightarrow R)$ be continuous, where (X) is compact.

2. Goal:

Show that (f) has a maximum and a minimum value on (X).

3. Image of (f):

Consider the image $(f(X) \subseteq R)$

Since (X) is compact and (f) is continuous, $(f(X))$ is compact in (R).

4. Compact Subset of (R):

A set in (R), a is considered compact if and only if it satisfies the conditions of being both closed and bounded (Perfekt, 1997).

5. Attaining Maximum and Minimum:

i. Being bounded, $(f(X))$ has a supremum and an infimum, say $(supsup\, f(X) = M)$ and $(infinf\, f(X) = m)$.

ii. Since $(f(X))$ is closed, $(M \in f(X))$ and$(m \in f(X))$

iii. Hence, there exist points $(x_M \in X)$ and $(x_m \in X)$ such that $(f(x_M) = M)$ and $(f(x_m) = m)$.

Therefore, (f) attains its maximum value (M) and minimum value (m) on (X).

This theorem has profound implications in analysis and optimization, ensuring that continuous functions on compact domains are well-behaved in terms of their extreme values (Jha & Triar, 2021).

Applications:

1. Optimization:

The Extreme Value Theorem is essential in optimization problems, guaranteeing the existence of optimal solutions for continuous functions over compact sets.

2. Economics:

In economics, the theorem ensures that utility functions, which are often continuous, achieve maximum utility over compact choice sets (Afuecheta et al., 2023).

3. Physics:

In physics, the theorem can be used to find equilibrium states in systems where potential energy, which is typically a continuous function, needs to be minimized.

4. Mathematical Analysis:

It is possible to demonstrate additional significant conclusions in the field of mathematical analysis with the assistance of the theorem. These results include the presence of solutions to differential equations and variational issues.

The interplay between compactness and continuity, highlighted by the Continuous Images of Compact Sets Theorem and the Extreme Value Theorem, is fundamental to many areas of mathematics, providing essential tools for both theoretical and applied studies.

3.2.3. Limit Point Compactness

Limit point compactness is an alternative way to define compactness, which specifically considers the limit points of sets.

A topological space (X) is said to be a limit point compact if every infinite subset of (X) has at least one limit point that belongs to (X).

A limit point, also known as an accumulation point, of a subset $(A \subset X)$ is a point $(x \in X)$ that satisfies the condition that any open neighborhood of (x) includes at least one point of (A) that is distinct from (x) itself.

Limit point compactness is another characterization of compactness, focusing on limit points of sets (Janssen & Marohn, 1994).

The following are properties of Limit Point Compactness:

1. Closed Set Property:

If (X) is the limit point compact, any closed subset of (X) is also the limit point compact. This follows because the limit points of any subset must lie within the closed subset.

2. Relation to Countable Compactness:

In countable compactness, a space is defined as having a finite subcollection that remains covered by any countable collection of open sets that covers the space. The terms "compactness," "countable compactness," and "limit point compactness" are interchangeable or synonymous in metric spaces (Ginsburg, 1975).

3. Bolzano-Weierstrass Property:

   Limit point compactness is often referred to as the Bolzano-Weierstrass property, as the Bolzano-Weierstrass Theorem implies that (R^n) is limit point compact.

a. **Proof of Limit Point Compactness in (R^n):**

The Bolzano-Weierstrass Theorem states that every limited infinite subset of (R^n) has a limit point. This implies that (R^n) is limit point compact (Horn, 1970).

b. Compact Implies Limit Point Compact:

i. Suppose (X) is compact. Let $(A \subseteq X)$ be an infinite subset.

ii. If (A) had no limit point in (X), then for each $(x \in X)$, there exists an open neighborhood (U_x) such that $(U_x \cap A)$ contains at most one point.

iii. The collection $(\{U_x\}_{x \in A})$ forms an open cover of (X).

iv. By compactness, there exists a finite subcover $(\{U_{x_1}, U_{x_2}, \ldots, U_{x_n}\})$.

v. Since each U_{x_i} contains at most one point of (A), this finite subcover can cover only finitely many points of (A), contradicting the assumption that (A) is infinite (Çakalli, 2008).

vi. Therefore, (A) must have a limit point in (X).

3.2.3.1. Relation to Sequential Compactness

There is a strong connection between limit point compactness and sequential compactness, particularly in metric spaces for example.

Theorem:

By definition, a space is considered to be compact in metric spaces only if it is sequentially compact.

Relation:

1. **Sequentially Compact** $(\backslash implies)$ **Limit Point Compact:**

i. Suppose (X) is sequentially compact. Let $(A \subseteq X)$ be an infinite subset.

ii. Consider any sequence $(\{a_n\}) in (A)$.

iii. By sequential compactness, there exists a subsequence $(\{a_{n_k}\})$ converging to some $(x \in X)$.

iv. Since $(\{a_{n_k}\})$ is a subsequence of $(\{a_n\})$, the point (x) is a limit point of (A).

v. Thus, (X) is a limit point compact.

2. **Limit Point Compact** \$(\backslash implies)$ **Sequentially Compact:**

i. Suppose (X) is limit point compact. Let $(\{x_n\})$ be any sequence in (X).

ii. If $(\{x_n\})$ is finite, it trivially has a convergent subsequence.

iii. If $(\{x_n\})$ is infinite, consider its set of values, say (A).

iv. By limit point compactness, (A) has a limit point $(x \in X)$.

v. Construct a subsequence $(\{x_{n_k}\})$ converging to (x).

vi. Therefore, (X) is sequentially compact.

3.2.3.2. Implications and Applications

1. Analysis:

   Limit point compactness is used in proving the existence of solutions to various problems, particularly in real analysis and functional analysis (Conidis, 2012).

2. Function Spaces:

   In spaces of continuous functions, limit point compactness aids in establishing properties like the Arzelà-Ascoli Theorem, which is crucial for understanding compactness in function spaces.

3. Topological Consequences

   The concept of limit point compactness helps in exploring deeper topological properties and their implications for other types of compactness, providing a rich framework for theoretical investigations.

   By understanding limit point compactness and its relationship with sequential compactness, valuable insights are gained into the structure and properties of compact spaces, which are pivotal in both pure and applied mathematics (Michel et al., 2007).

3.2.4. Applications in Analysis

3.2.4.1. Compact Operators

In the field of functional analysis, compact operators are a very important component, particularly when it comes to the investigation of vector spaces with infinite dimensions.

An operator $(T : X \rightarrow Y)$ between two Banach spaces (X) and $(\ )$ is called compact if it maps bounded sets in (X) to relatively compact sets in (Y). This means that the closure of $(T(B))$ is compact in (Y) for every bounded set $(B \subseteq X)$.

The following are the properties of Compact Operators (Axler & Zheng, 1998):

1. Continuity:

Every compact operator is continuous.

2. Spectrum:

On an infinite-dimensional Banach space, the spectrum of a compact operator is countable and can only build up to zero.

3. Approximation:

Finite-rank operators may often approach compact operators, which makes calculations and analysis easier.

Following are the examples:

1. Integral Operators:

Consider the operator $\left(T : L^2([a,b]) \to L^2([a,b])\right)$ defined by $\left((Tf)(x) = \int_a^b K(x,y) f(y) dy \right)$, where (K) is a continuous kernel. (T) is compact due to the integral's smoothing effect (Calderón & Zygmund, 1957).

2. Finite-Rank Operators:

Operators that map any element of (X) to a finite-dimensional subspace of (Y) are trivially compact.

Theorems Involving Compact Operators:

1. **Fredholm Alternative:**

For a compact operator (T), either $\left((I-T)x = 0\right)$ has only the trivial solution, or $\left((I-T)x = y\right)$ has a solution for every $(y \in X)$.

2. **Spectral Theory:**

Compact operators have a discrete spectrum, and their eigenvalues (excluding zero) have finite multiplicity (Koppelman, 1964).

3.2.4.2. Arzelà-Ascoli Theorem

A characterization of relatively compact subsets of the space of continuous functions on a compact set (K) is given by the Arzelà-Ascoli Theorem.

Statement (Antonescu & Christensen, 2004):

A subset $(F \subseteq C(K))$ is relatively compact if and only if it is uniformly bounded and equicontinuous.

1. Uniformly Bounded: (F) is uniformly bounded if there exists $(M>0)$ such that $(|f(x)|\leq M)$ for all $(f\in F)$ and $(x\in K)$.
2. Equicontinuous: (F) is equicontinuous if for every $(\epsilon>0)$, there exists $(\delta>0)$ such that $(|f(x)-f(y)|<\epsilon)$ for all $(f\in F)$ and $(x,y\in K)$ with $(d(x,y)<\delta)$.

Proof:

1. **Necessity:**

   Assume (F) is relatively compact. Uniform convergence implies uniform boundedness and equicontinuity.
2. **Sufficiency:**

   Assume (F) is uniformly bounded and equicontinuous. Using the equicontinuity, construct a sequence of finite (ϵ). Employ the diagonal argument to extract a uniformly convergent subsequence, thus proving relative compactness (Krukowski, 2018).

Applications:

1. Existence of Solutions:

   Differential equation solutions are shown to exist using the Arzelà-Ascoli Theorem. A convergent subsequence whose limit is a solution may be extracted by demonstrating the relative compactness of an estimated solution sequence.
2. Functional Analysis:

   In functional analysis, the theorem is employed to characterize compact operators. For instance, if $(T:C(K)\to C(K))is$ an operator that maps bounded sets to relatively compact sets, (T) can be shown to be compact using Arzelà-Ascoli.
3. Probability Theory:

   The theorem aids in the study of distributional convergence and the establishment of the tightness of probability measures in probability theory (Klein, 1986).

Examples:

1. Integral Equations:

   Consider the integral equation $\left(f(x)=g(x)+\int_a^b K(x,y)f(y)dy\right)$. The Arzelà-Ascoli Theorem can be used to show that the operator defined by the integral is compact, ensuring the existence of solutions under certain conditions (Hsiao & Wendland, 2008).

2. Differential Equations:

   In the realm of boundary value problems, this theorem is helpful for confirming the presence of solutions by demonstrating that the set of possible solutions is sufficiently compact. This has the effect of establishing that solutions do exist.

When doing analysis, compactness is of the utmost importance, particularly when taking into consideration compact operators and the Arzelà-Ascoli Theorem. Approximations and spectral analysis are made possible by compact operators, which make it possible to investigate spaces that have an unlimited number of dimensions. In the meanwhile, the Arzelà-Ascoli Theorem performs the role of a powerful instrument for gaining an understanding of the properties of function spaces. One of the most important concepts in advanced mathematical analysis is compactness (Gronwall, 1919).

3.3. CONNECTED AND PATH-CONNECTED SPACES

3.3.1. Definition and Basic Properties

A topological space (X) is said to be "connected" if it cannot be divided into two disjoint non-empty open subsets. In other words, there are no two open sets (U) and (V) in (X) such that $(X = U \cup V), (U \cap V = \varnothing)$, and both (U) and (V) are non-empty.

A space is connected if there is no "gap" that divides the space into two parts. Imagine a rope; if the rope is whole and uncut, it represents a connected space. If the rope is cut into two pieces, it represents a disconnected space (Alimov & Bednov, 2021).

3.3.2. Examples of Connected Spaces

3.3.2.1. The Real Line (R)*:*

The set of real numbers with the standard topology is a classic example of a connected space. Intuitively, this is because there is no way to partition (R) into two non-empty open sets without leaving gaps. Formally, if (R) were to be split into two non-empty open sets (U) and (V), there would be points $(a \in U)$ and $(b \in V)$ such that $(a<b)$. Due to the density of the real numbers, any interval between (a) and (b) would also belong to either (U) or (V), leading to a contradiction.

Intervals in (R):

Any interval in (R), whether open, closed, or half-open, is connected. For example, the interval $([0,1])$ is connected because if it were split into two non-empty open sets, there would be points within $((0,1))$ that could not be separated cleanly without breaking the continuity of the interval.

The Complex Plane (C):

Like the real line, the complex plane with the standard topology is connected. There is no way to partition the entire plane into two non-empty open subsets without leaving some regions unexplored or unconnected (Motzkin, 1951).

3.3.2.2. Topologist's Sine Curve

The topologist's sine curve is another example of a connected space. It consists of the graph of the function $\left(y = sinsin\left(\frac{1}{x}\right)\right)$ for $(0<x\leq 1)$ together with the segment on the (y)–axis where $(x=0)$ and $(-1\leq y\leq 1)$. Despite its complicated appearance, this space is connected because it is impossible to partition it into two non-empty open sets (Chinn & Steenrod, 1966).

3.3.2.3. Subsets of (R^n)

For instance, if $(f:R\rightarrow R^n)$ is a continuous function, then the image of (mathbb{R}) under (f) is connected. This is because the continuous image of a connected space retains the property of connectedness.

3.3.3. Basic Properties of Connected Spaces

Preservation under Continuous Mappings:

If $(f:X\rightarrow Y)$ is a continuous function and (X) is connected, then the image $(f(X))$ is also connected. This property is pivotal in various fields of mathematics, ensuring that connected structures remain intact under continuous transformations (Neumann-Lara & Wilson, 1986).

3.3.3.1. Connected Components

Every topological space (X) can be decomposed into maximal connected subsets called connected components. Each connected component is closed in (X).

3.3.3.2. Intermediate Value Theorem

The Intermediate Value Theorem is a direct consequence of connectedness in the real line. If $(f:[a,b]\to R)$ is continuous and $(f(a)\neq f(b))$, then for any value (c) between $(f(a))$ and $(f(b))$, there exists some $(x\in[a,b])$ such that $(f(x)=c)$. This theorem illustrates how connectedness restricts the behavior of continuous functions.

3.3.3.3. Subspace Connectedness

A subspace $(A\subseteq X)$ is connected if and only if (A) cannot be partitioned into two non-empty disjoint open subsets in the subspace topology. This means that connectedness is preserved when considering subspaces of connected spaces (Alas et al., 1999).

3.3.3.4. Union of Connected Sets

A collection of linked sets is connected when its union has a common point. This property is useful in constructing larger connected spaces from smaller ones by ensuring overlapping regions maintain overall connectedness (Lianqiang et al., 2017).

Understanding connectedness is fundamental in topology, as it provides insights into the structure and behavior of spaces under various transformations. Connected spaces serve as the building blocks for more complex topological constructs and have applications in analysis, geometry, and beyond (Jayanti & Tarjan, 2016) (Figure 3.3).

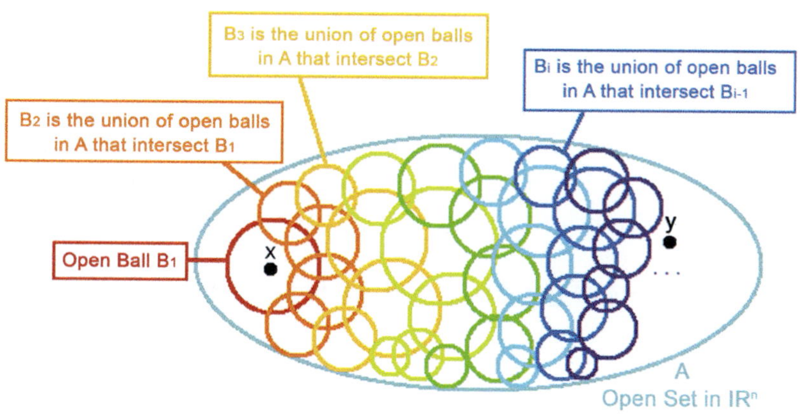

Figure 3.3. Illustration of the Unions of the connected spaces (Source: Math Online, Creative Commons License)

3.3.4. Path-Connected Spaces

A topological space (X) is said to be "path-connected" if, for any two points $(x, y \in X)$, there exists a continuous map $(f:[0,1] \to X)$ such that $(f(0)=x)$ and $(f(1)=y)$. This map (f) is called a "path" from (x) to (y) (Figure 3.4).

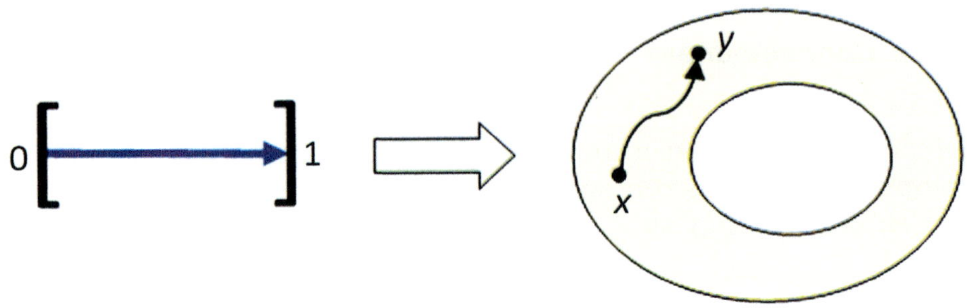

Figure 3.4. Illustration of path connected spaces (Source: Math Strek, Creative Commons License)

3.3.4.1. Examples of Path-Connected Spaces

1. **The Real Line $(\boldsymbol{R})$:**

The real line (R) is path-connected because for any two points $(x, y \in R)$, the linear path $(f(t)=(1-t)x+ty)$ for $(t \in [0,1])$ is a continuous map from (x) to (y).

2. **Euclidean Space $(\boldsymbol{R^n})$:**

Similar to the real line, Euclidean spaces (R^n) are path-connected. Given two points $(x, y \in R^n)$, *the* path $(f(t)=(1-t)x+ty)$ for $(t \in [0,1])$ is a continuous map from (x) to (y).

3. Circles and Spheres:

A circle (S^1) and a sphere (S^n) are path-connected. For instance, on a circle, any two points can be connected by an arc along the circumference. Similarly, on a sphere, any two points can be joined by a geodesic path (Miteva & Lazarova, 2018).

4. **Connected Subspaces of (R^n):**

Many connected subspaces of (R^n) are also path-connected. For example, an interval

$([a,b] \subset R)$ is path-connected because for any $(x, y \in [a,b])$, the path $(f(t) = (1-t)x + ty)$ lies entirely within $([a,b])$.

5. Graphs:

In graph theory, a connected graph is path-connected. Any two vertices in a connected graph can be connected by a sequence of edges, which forms a path in the graph.

3.3.4.2. Relation to Connected Spaces

While path-connectedness implies a stronger form of connectedness, the two concepts are closely related but not equivalent (Macbeath, 1964).

Path-Connectedness Implies Connectedness:

If a space (X) is path-connected, then it is connected. This is because if (X) could be partitioned into two disjoint non-empty open sets (U) and (V), it would be impossible to find a path from a point in (U) to a point in (V), contradicting the path-connectedness of (X).

Proof (Monteiro, 1987):

Assume (X) is path-connected but not connected. Then there exist disjoint non-empty open sets (U) and (V) such that $(X = U \cup V)$. Take $(x \in U)$ and $(y \in V)$. Since (X) is path-connected, there exists a continuous path $(f : [0,1] \rightarrow X)$ from (x) to (y). The preimage $(f^{-1}(U))$ and $(f^{-1}(V))$ form a partition of $([0,1])$ into non-empty disjoint open sets, contradicting the connectedness of $([0,1])$. Thus, (X) must be connected (Brodskiy et al., 2008).

Connectedness Does Not Imply Path-Connectedness:

A space can be connected but not path-connected. A classic example is the "topologist's sine curve." This space consists of the graph of $(y \quad sinsin(1/x))$ for $(0 < x \leq 1)$ together with the vertical segment on the (y)–axis from $(\quad)$ to (1). The space is connected but not path-connected because there is no continuous path connecting points on the vertical segment to points on the sine curve.

Key Differences:

1. Intermediate Value Theorem: In connected spaces, it holds for continuous functions. In path-connected spaces, this extends to paths (Macbeath, 1964).
2. Local Properties: Path-connectedness is a local property as small neighborhoods around points can determine path-connectedness, whereas connectedness can be more global.
3. Complex Structures: Spaces with complex local structures can be connected without being path-connected.

Understanding path-connected spaces and their relationship to connected spaces is crucial in topology. Path-connectedness provides a stronger condition that ensures more robust connectivity properties, facilitating analysis and application in various mathematical contexts. However, the distinction between the two concepts highlights the nuanced and rich structure that topological spaces can exhibit (Fischer et al., 2011).

3.3.5. Components and Path Components

3.3.5.1. Connected Components

Connected components are maximal connected subsets within a topological space. Formally, a "connected component" of a topological space (X) is a connected subset (C) that is not properly contained in any other connected subset of (X).

The following are the properties of Connected Components (Albert & Youngs, 1942):

1. Uniqueness: Each point in (X) belongs to exactly one connected component. The connected components of (X) form a partition of (X).
2. Closed Sets: Connected components are always closed in (X). If (C) is a connected component of (X), then (C) is closed under the subspace topology.
3. Disjointness: Different connected components are disjoint. If (C_1) and (C_2) are two distinct connected components, then $(C_1 \cap C_2 = \varnothing)$.
4. Local Connectivity: If (X) is locally connected, then the connected components of (X) are also open.

Examples:

1. Real Line with a Point Removed: Consider $(R \setminus \{0\})$. The space is divided into two connected components: $((-\infty, 0))$ and $((0, \infty))$.
2. Discrete Spaces: In a discrete space, every singleton set is a connected component because no larger subset is connected (Ekici & Noiri, 2008).

3.3.5.2. Path Components

Path components, or path-connected components, are subsets of a topological space where any two points can be joined by a continuous path lying entirely within the subset (Conlon, 1964). A topological space (X) is "path-connected" if, for every pair of points $(x, y \in X)$, there exists a continuous map $(f : [0,1] \rightarrow X)$ such that $(f(0) = x)$ and $(f(1) = y)$.

The following are the properties of Path Components (Virk & Zastrow, 2017):

1. Inclusion: Every path-connected space is connected, but not every connected space is path-connected.
2. Maximality: Path components are maximal path-connected subsets of (X). A path component cannot be extended by adding more points without losing the path-connected property (Witten, 2016).
3. Uniqueness: Each point in (X) belongs to exactly one path component. Path components form a partition of (X).
4. Openness and Closedness: In general, path components need not be open or closed (Bagchi, 2021) (Figure 3.5).

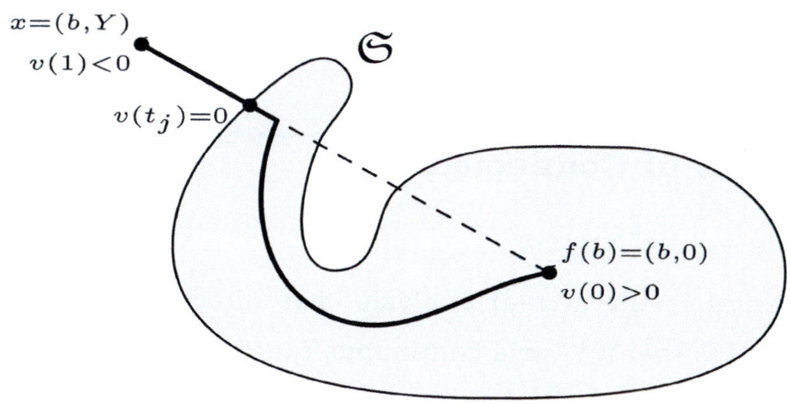

Figure 3.5. Illustration of the path-connected components of V (b) ∩ R (Source: Yves Balasko, Creative Commons License)

Examples:

1. Real Line (R): The real line is path-connected because any two points can be connected by a straight line, which is a continuous path.
2. Circle (S^1): The circle is path-connected as any two points can be joined by a path along the circumference.
3. Topologist's Sine Curve: It is connected but not path-connected. The part $(y = sin\ sin\ \left(\frac{1}{x}\right))$ is connected but lacks a path component due to the oscillatory behavior near $(x=0)$.

Relation between Connected and Path-Connected Components:

Path components are always contained within connected components, but they are not necessarily the same. In spaces where each connected component is also path-connected, the concepts coincide. However, in more complex spaces, such as those with intricate local structures, connected components may contain multiple path components (Maciejewski & Fox, 1993).

Key Differences:

1. General Spaces: In general topological spaces, connected components, and path components can differ significantly.
2. Locally Path-Connected Spaces: If a space is locally path-connected, its connected components and path components coincide, making the space simpler to analyze (McCammon & Hollinger, 2021).

Connected components capture the broader sense of "wholeness" in a space, while path components provide a more specific criterion based on continuous paths. These concepts play pivotal roles in the analysis and classification of topological spaces, particularly in determining the structure and behavior of various types of spaces in mathematical and applied contexts (Zastrow, 1994).

3.3.6. Applications of Connectedness

3.3.6.1. Intermediate Value Theorem

This is a fundamental result in real analysis that directly relies on the concept of connectedness. It states that if (f) is a continuous function on a closed interval $([a,b])$ and (d) is any value between $(f(a))$ and $(f(b))$, then there exists some $(c \in [a,b])$ such that $(f(c)=d)$.

Formal Statement (Ghosh et al., 2022):

If $(f:[a,b]\to R)$ is continuous and $(f(a)\leq d\leq f(b))$ (or $(f(b)\leq d\leq f(a))$), then there exists $(c\in[a,b])$ such that $(f(c)=d)$.

Proof:

Assume without loss of generality that $(f(a)\leq d\leq f(b))$. Define the sets:

$$\left[A=\{x\in[a,b]\mid f(x)\leq d\}\right]$$

$$\left[B=\{x\in[a,b]\mid f(x)\geq d\}\right]$$

Both sets (A) and (B) are closed and $([a,b]=A\cup B)$. Since (f) is continuous, (A) and (B) are connected, and thus they cannot partition $([a,b])$ into disjoint non-empty sets (Russ, 1980). Hence, there must be some $(c\in[a,b])$ where $(f(c)=d)$.

3.3.6.2. Connectedness in Real Analysis

Connectedness plays a crucial role in various theorems and concepts in real analysis, ensuring the integrity of functions and spaces (Nguyen et al., 2022).

1. Continuous Functions Preserve Connectedness:

   If $(f:X\to Y)$ is a continuous function and (X) is connected, then $(f(X))$ is also connected. This is used to demonstrate the behavior of continuous functions over connected domains, ensuring that no "gaps" form in the image of the function (Zou et al., 2024).

   Example: Consider the function $(f(x)=x^2)$ on (R). Since (R) is connected and (f) is continuous, the image $(f(R)=[0,\infty))$ *is* connected.
2. Extreme Value Theorem:

   It states that if (f) is continuous on a compact interval $([a,b])$, then (f) attains both its maximum and minimum values on $([a,b])$. The connectedness of $([a,b])$ ensures that (f) cannot "escape" these bounds without violating continuity (Mensi et al., 2023).

3. *Connected Subsets of* (R):

Understanding connected subsets of (R) helps in the analysis of intervals and other real-valued functions. Intervals are the simplest connected subsets, providing a foundation for more complex constructions.

Example: Let $(X = \{x \in R \mid 0 \le x < 1\} \cup \{2\}).(X)$ is not connected because it can be partitioned into $([0,1))$ and $(\{2\})$.

Connectedness is a fundamental concept in real analysis, underpinning key theorems and ensuring the coherent behavior of continuous functions (Liow & Huang, 2018).

3.4. APPLICATIONS OF COMPACTNESS AND CONNECTEDNESS

3.4.1. Compactness in Function Spaces

1. Compact-Open Topology
2. Applications in Functional Analysis

3.4.1.1. Compact-Open Topology

The compact-open topology is a fundamental topology on the set of continuous functions between two topological spaces, particularly useful in the study of function spaces. Given topological spaces (X) and (Y), the set of all continuous functions from (X) to (Y) is denoted by $(C(X,Y))$:

$$\left[W(K,U)=\{f \in C(X,Y) \mid f(K) \subseteq U\}\right]$$

Where $(K \subseteq X)$ is compact and $(U \subseteq Y)$ is open. The sub-basis elements $(W(K,U))$ capture the idea that a function (f) in $(C(X,Y))$ maps the compact set (K) into the open set (U).

Properties (O'Meara, 1971):

1. Continuity: A sequence of functions $(\{f_n\})$ converges to (f) in the compact-open topology if and only if for every compact set $(K \subseteq X)$ and every open set $(U \subseteq Y)$, there exists an (N) such that for all $(n \geq N), (f_n(K) \subseteq U)$.
2. Locally Compact Spaces: If (X) is locally compact and Hausdorff, the compact-open topology on $(C(X,Y))$ aligns well with the natural topological structure of (X).

3.4.1.2. Applications in Functional Analysis

Compactness plays a critical role in various theorems and concepts in functional analysis, influencing the behavior and properties of function spaces.

1. Banach-Alaoglu Theorem:

It states that the closed unit ball in the dual space of a normed vector space is compact in the weak topology. This result has significant implications in functional analysis, particularly in the study of Banach spaces and their duals (Guo, 2008).

Applications:

i. Optimization Problems: It is used in optimization problems to show the existence of optimal solutions by compactifying the set of potential solutions in the weak topology.

ii. Functional Spaces: In the study of functional spaces, the theorem helps in understanding the compactness properties of functionals and their implications on the behavior of sequences in these spaces.

Compactness in function spaces, particularly through the compact-open topology, provides a robust framework for analyzing continuous functions and their properties. The applications in functional analysis, highlighted by the Arzelà-Ascoli and Banach-Alaoglu theorems, demonstrate the profound impact of compactness on the study and application of functional spaces. Understanding these concepts is crucial for advancing in areas like differential equations, optimization, and the broader field of functional analysis (Brattka, 2008).

3.4.2. Compactness and Convergence

3.4.2.1. Convergence of Functions

In the context of function spaces, convergence can be defined in several ways, depending on the topology considered. The most common types are:

Pointwise Convergence:

A sequence of functions $(\{f_n\})$ converges pointwise to a function (f) if for every $(x \in X)$ and every $(\epsilon > 0)$ there exists an (N) such that for all $(n \geq N), (|f_n(x) - f(x)| < \epsilon)$. Formally,

$$\left[f_n(x) = f(x) \quad \text{for all } x \in X. \right]$$

Properties:

i. Dependence on (x): Pointwise convergence allows different rates of convergence at different points in the domain.

ii. Compactness and Pointwise Convergence: If $(\{f_n\})$ is a sequence of continuous functions on a compact set $(K \subset X)$ converging pointwise to (f), (f) need not be continuous (Brézis & Lieb, 1983).

3.4.2.2. Uniform Convergence

Uniform convergence provides a stronger form of convergence that ensures the uniformity of the rate of convergence across the entire domain.

A sequence of functions $(\{f_n\})$ converges uniformly to a function (f) on a set (X) if for every $(\epsilon > 0)$, there exists an (N) such that for all $(n \geq N)$ and all $(x \in X)$,

$$[|f_n(x) - f(x)| < \epsilon.]$$

In other words,

$$\left[\left| f_n(x) - f(x) \right| = 0. \right]$$

Properties:

1. Preservation of Continuity: If $(\{f_n\})$ is a sequence of continuous functions on a set (X) and (f_n) converges uniformly to (f), then (f) is continuous.
2. Compactness and Uniform Convergence: Compactness plays a significant role in ensuring uniform convergence.

Applications of Uniform Convergence:

1. Integration and Differentiation: Uniform convergence allows the interchange of limits with integrals and derivatives. If $(f_n \to f)$ uniformly and each (f_n) is integrable or differentiable (Baltes et al., 1980), then

$$\left[\int f_n(x)dx = \int f_n(x)\, dx \right]$$

and

$$\left[(f_n(x))' = f_n'(x). \right]$$

2. *Equicontinuity and Arzelà-Ascoli Theorem:* It states the relative compactness of function sequences. It states that a family of functions $(F \subset C(X, Y))$ is relatively compact if and only if it is uniformly bounded and equicontinuous. Uniform convergence is a key aspect of applying this theorem.

Compactness and convergence are intertwined concepts in analysis, especially in the context of function spaces. Uniform convergence ensures robust properties such as the preservation of continuity and the interchangeability of limits with integrals and derivatives. Understanding these types of convergence and their implications is essential for advanced studies in real and functional analysis.

3.4.3. Connectedness in Topological Groups

3.4.3.1. Connected Topological Groups

A topological group is a group (G) equipped with a topology such that the group operations (multiplication and inversion) are continuous with respect to that topology. Connectedness in topological groups provides insight into the structure and behavior of these groups.

A topological group (G) is "connected" if it cannot be partitioned into two disjoint non-empty open subsets (Keesling, 1974) (Figure 3.6).

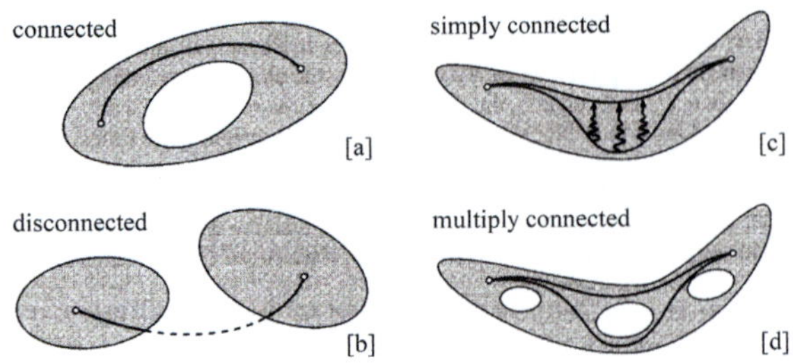

Figure 3.6. Illustrations of the Connected Topological Groups (Source: Stack Exchange, Creative Commons License)

Examples:

1. The Real Numbers $(\boldsymbol{R})$ with Addition:

   The group $((R,+))$ is connected because the real line cannot be partitioned into disjoint non-empty open subsets.

2. The Circle Group (S^1):

   The group of complex numbers with absolute value 1, $(S^1 = \{z \in C \mid |z| = 1\})$, is connected in its standard topology, as it forms a continuous loop without breaks.

3. Lie Groups:

   Many Lie groups, such as the special orthogonal group $(SO(n))$ and the special unitary group $(SU(n))$, are connected. These groups are both topological groups and smooth manifolds, providing rich examples of connected topological groups.

Properties:

1. Product of Connected Groups: The product of connected topological groups is connected. If (G) and (H) are connected topological groups, then their direct product $(G \times H)$ is also connected (Hofmann & Morris, 2007).
2. Connected Components: The connected components of a topological group are co-sets of the identity component (G_e).

3.4.3.2. Applications in Group Theory

1. Homomorphisms and Connectedness:

Homomorphisms between topological groups can preserve connectedness. If $(\phi{:}G \to H)$ is a continuous homomorphism and (G) is connected, then the image $(\phi(G))$ is connected.

Example: Consider the exponential map $(exp : R \to S^1)$ is a homomorphism from the additive group of real numbers to the multiplicative group of complex numbers with unit magnitude. Since (R) is connected, $(expexp(R) = S^1)$ is also connected.

2. Topology and Group Structure:

Understanding the connectedness of a topological group can provide insights into its algebraic structure. For instance, connected Lie groups often have simpler and more tractable structures compared to their disconnected counterparts.

3. Representation Theory:

Connectedness plays a role in the representation theory of topological groups. For connected Lie groups, the study of representations often simplifies, as connected components do not introduce additional complexity.

Example: The representation theory of $(SO(3))$, the group of rotations in three dimensions, leverages its connectedness to understand how the group acts on various vector spaces and manifolds.

4. Topological Dynamics:

It is possible for the connectedness of the underlying group to have an effect on the behavior of group actions when topological dynamics are discussed. For instance, as compared to actions carried out by disconnected groups, the actions carried out by linked groups on topological spaces can display behaviors that are more consistent and continuous.

The concept of connectedness in topological groups is significant in both topology and group theory, and it is a foundational topic that has substantial ramifications. The structure of the groups, their representations, and the actions they take on other spaces are all easier to comprehend as a result of this. The study of connected topological groups helps to bridge the gap between algebraic properties and topological properties, resulting in a deeper comprehension of both of these areas of mathematics (Sell, 1967).

3.4.4. Advanced Applications

3.4.4.1. Compactness and Connectedness in Algebraic Topology

Algebraic topology is a branch of topology that makes use of techniques from abstract algebra in order to achieve its goal of studying topological spaces. Due to the fact that they serve as a basis for understanding topological systems that are becoming increasingly complex, compactness and connectedness are two of the most significant concepts in this field of research m(Janos & Rosenfeld, 1983).

1. Compactness in Algebraic Topology:

It is common practice to apply compactness when it comes to the investigation of theories about homology and cohomology. The compactness of a space, for instance, may make the calculation of the homology groups of that space easier to understand and perform.

For example, when computing simplicial homology, the compactness of a simplicial complex ensures that every open cover has a finite subcover. This is a guarantee that every open cover has a finite subcover. Because of this, defining simplices and dealing with them is an absolutely necessary requirement (Sizemore et al., 2019).

2. Connectedness in Algebraic Topology:

Example: The basic group $(\pi_1(X))$ of a linked space (X) offers valuable information about the loops present in (X). In algebraic topology, the basic group remains unchanged under homotopy for path-connected spaces. Homotopy is a crucial notion in this field.

Applications:

1. Fundamental Group: The fundamental group (π_1) measures the connectedness of loops in a space. For a connected space, (π_1) helps classify covering spaces and provides information about the space's topology.

2. Homology and Cohomology: Compactness is used to define and compute homology and cohomology groups. These groups are essential in distinguishing topological spaces and understanding their properties.

3.4.4.2. Use in Modern Research and Applications

Modern research in topology and its applications often leverages the concepts of compactness and connectedness.

1. Topological Data Analysis:

In order to investigate the structure of data, topological data analysis (TDA) makes use of the ideas that are derived from algebraic topology. When conducting an examination of persistent homology, which is a technique that identifies enduring patterns that are present in data sets of varying sizes, compactness and connectedness are elements that are utilized.

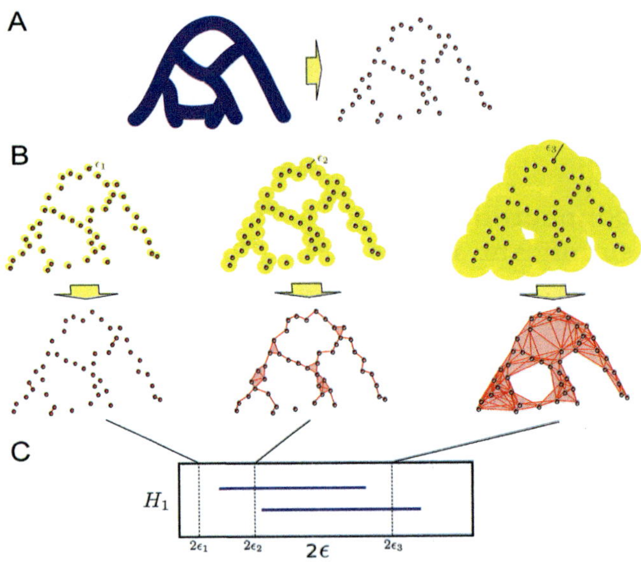

Figure 3.7. Data analysis of topological data. (A) From a finite number of sampled points, topological data analysis seeks to deduce the topological properties (such as loops, voids, etc.) of an unknown space. (B) Using balls of radius ϵ centered on the sampled points, on the sampled points, persistent homology, a TDA technique, constructs simplicial complexes—generalizations of networks that incorporate higher dimensional elements like triangles and tetraheadra. If the respective balls overlap, then the points in the simplicial complex are connected. The Vitoris-Rips complex is the name given to this structure. The evolution of the topological characteristics of Vietoris-Rips complexes with ϵ is monitored using persistent homology. (C) Barcodes are an appropriate way to depict persistent homology. Every barcode interval denotes the range of ρ that a specific topological feature (such as a loop) is found in the inferred topology (Source: Pablo G. Camara, Creative Commons License)

For instance, in TDA, the compactness of data points guarantees meaningful interpretations of persistent properties, whereas connectedness assists in the identification of clusters and components in high-dimensional data (Chow, 1959).

2. Quantum Field Theory and String Theory:

The concepts of compactness and connectedness are utilized in the field of theoretical physics, namely in the study of space-time and string theory. Compact manifolds are utilized in string theory for the purpose of modeling compact dimensions, which in turn has an effect on the behavior of physical theories.

As an illustration, compactification in string theory refers to the process of reducing the dimensions of space-time by the use of compact linked manifolds, which ultimately results in the utilization of rich mathematical structures and physical predictions.

3. Complex Networks:

Understanding the structure and dynamics of networks is essential for the study of complex networks, and connectivity is an essential component of this study. For the purpose of analyzing social, biological, and technical networks, concepts such as connected components and network compactness are taken into consideration.

As an illustration, in the field of social network analysis, the components of a network that are connected indicate the subgroups as well as the general connectivity of the network. This provides insights into the dissemination of information and the impact that is exerted.

In algebraic topology, compactness and connectedness are important concepts that provide essential tools for understanding and studying topological spaces due to their ability to provide essential tools. The examples of their applications in contemporary research, which range from topological data analysis to theoretical physics and complex networks, demonstrate the significance and diversity of these mathematical concepts in both pure and practical mathematics (Serrano et al., 2020).

ACTIVITY 3.1.

Explore the concept of path-connected spaces and investigate examples where path-connectedness differs from connectedness, emphasizing the role of paths in connectivity.

SUMMARY

Chapter 3 delves into the fundamental concepts of compactness and connectedness within the realm of topology. The chapter begins with an introduction to compact spaces, providing a rigorous definition and exploring their basic properties. It highlights the importance of the Heine-Borel Theorem, a cornerstone in topology, and illustrates its applications in various contexts. The chapter also extends the notion of compactness to product spaces and discusses sequential compactness, emphasizing its significance in metric spaces.

Further, the chapter explores the characterization of compactness in metric spaces, examining its relationship with continuous functions and limit point compactness. This section includes practical applications in analysis, demonstrating the utility of compactness in understanding the behavior of functions and sequences.

The latter part of the chapter focuses on connected and path-connected spaces, defining these concepts and elucidating their basic properties. Examples of connected spaces are provided to enhance understanding, followed by a discussion on path-connected spaces, components, and path components. The chapter concludes with applications of connectedness in topological groups and other advanced topics, underscoring the relevance of these concepts in broader mathematical contexts.

Through rigorous definitions, theorems, and applications, this chapter aims to equip the reader with a deep understanding of compactness and connectedness, two pivotal concepts in topology that have profound implications in various areas of mathematics and its applications.

MULTIPLE CHOICE QUESTIONS

1. **In a compact space, which of the following is true?**
   a. Every sequence has a convergent subsequence.
   b. Every open cover has a finite subcover.
   c. Every closed cover has a finite subcover.
   d. Every sequence converges.
2. **The Heine-Borel Theorem applies to:**
   a. Any metric space
   b. Any topological space
   c. Euclidean space
   d. Any connected space
3. **Sequential compactness implies:**
   a. Every sequence has a convergent subsequence.
   b. Every sequence converges.
   c. Every open cover has a finite subcover.

d. The space is connected.

4. A path-connected space is:

a. Always compact

b. Always connected

c. Always sequentially compact

d. Always locally compact

5. Compactness in product spaces is guaranteed by:

a. Heine-Borel Theorem

b. Tychonoff's Theorem

c. Bolzano-Weierstrass Theorem

d. Baire Category Theorem

6. In a connected space, which of the following is impossible?

a. The space can be divided into two non-empty disjoint open sets.

b. The space is path-connected.

c. The space is compact.

d. The space is locally compact.

7. An example of a limit point compact space is:

a. R with the standard topology

b. Q with the standard topology

c. The closed interval [0, 1] in R

d. R^2 with the standard topology

REVIEW QUESTIONS

1. Explain the definition of a compact space and provide an example.
2. Describe the Heine-Borel Theorem and its significance in topology.
3. What are the basic properties of compact spaces?
4. How does the concept of sequential compactness differ from general compactness?
5. Discuss the relationship between compactness and continuous functions in metric spaces.
6. Define connected and path-connected spaces and give examples of each.
7. Explain the applications of connectedness in topological groups.

Answers to Multiple Questions

1. (b) 2. (c) 3. (a) 4. (b) 5. (b)

6. (a) 7. (c)

REFERENCES

1. Afuecheta, E., Utazi, C., Ranganai, E., & Nnanatu, C. (2023). An Application of extreme value theory for measuring Financial Risk in BRICS Economies. *Annals of Data Science*, *10*(2), 251–290.
2. Alas, O. T., Tkačenko, M. G., Tkachuk, V. V., & Wilson, R. G. (1999). Connectedness and local connectedness of topological groups and extensions. *Commentationes Mathematicae Universitatis Carolinae*, *40*(4), 735–753.
3. Albert, G. E., & Youngs, J. W. T. (1942). The structure of locally connected topological spaces. *Transactions of the American Mathematical Society*, *51*, 637–654.
4. Alimov, A. R., & Bednov, B. B. (2021). Monotone path-connectedness of Chebyshev sets in three-dimensional spaces. *Sbornik: Mathematics*, *212*(5), 636.
5. Al-shami, T. M., Mhemdi, A., Abu-Gdairi, R., & El-Shafei, M. E. (2023). Compactness and connectedness via the class of soft somewhat open sets. *AIMS math*, *8*(1), 815–840.
6. Antonescu, C., & Christensen, E. (2004). Metrics on group C∗-algebras and a non-commutative Arzelà–Ascoli theorem. *Journal of Functional Analysis*, *214*(2), 247–259.
7. Ariza, J. G., Giraldez, I., Sanchez-Rodas, D., & Morales, E. (2000). Selectivity assessment of a sequential extraction procedure for metal mobility characterization using model phases. *Talanta*, *52*(3), 545–554.
8. Axler, S., & Zheng, D. (1998). Compact operators via the Berezin transform. *Indiana University Mathematics Journal*, 5, 387–400.
9. Azarpanah, F., Paimann, M., & Salehi, A. R. (2015). Compactness, connectedness and countability properties of C (X) with the r-topology. *Acta Mathematica Hungarica*, *146*, 265–284.
10. Bagchi, S. (2021). Connected Fundamental Groups and Homotopy Contacts in Fibered Topological (C, R) Space. *Symmetry*, *13*(3), 500.
11. Baltes, P. B., Cornelius, S. W., Spiro, A., Nesselroade, J. R., & Willis, S. L. (1980). Integration versus differentiation of fluid/crytallized intelligence in old age. *Developmental Psychology*, *16*(6), 625
12. Barone, J., & Novikoff, A. (1978). A history of the axiomatic formulation of probability from Borel to Kolmogorov: Part I. *Archive for history of exact sciences*, *18*, 123–190.
13. Bekkali, M., Bonnet, R., & Rubin, M. (1992). Compact interval spaces in which all closed subsets are homeomorphic to clopen ones, II: To the memory of Ernest corominas (1913–1992). Order, 9, 177–200.
14. Brattka, V. (2008). Borel complexity and computability of the Hahn–Banach theorem. *Archive for Mathematical Logic*, *46*(7), 547–564.
15. Brézis, H., & Lieb, E. (1983). A relation between pointwise convergence of functions and convergence of functionals. *Proceedings of the American Mathematical Society*, *88*(3), 486–490.
16. Bridges, D. S. (1978). On continuous mappings between locally compact metric spaces. *Bulletin of the London Mathematical Society*, *10*(2), 201–208.

17. Brodskiy, N., Dydak, J., Labuz, B., & Mitra, A. (2008). Covering maps for locally path-connected spaces. *arXiv preprint arXiv:0801.4967*, 456–460.

18. Bugajewski, D., & Gulgowski, J. (2020). On the characterization of compactness in the space of functions of bounded variation in the sense of Jordan. *Journal of Mathematical Analysis and Applications*, *484*(2), 123752.

19. Cai, Y., & Daskalakis, C. (2015). Extreme value theorems for optimal multidimensional pricing. *Games and Economic Behavior*, *92*, 266–305.

20. Çakalli, H. (2008). Sequential definitions of compactness. *Applied Mathematics Letters*, *21*(6), 594–598.

21. Calderón, A. P., & Zygmund, A. (1957). Singular integral operators and differential equations. In *Selected Papers of Antoni Zygmund*. Dordrecht: Springer Netherlands. (Vol. 3, pp. 257–277).

22. Cederquist, J., & Negri, S. (1995, June). A constructive proof of the Heine-Borel covering theorem for formal reals. In *International Workshop on Types for Proofs and Programs*. Berlin, Heidelberg: Springer Berlin Heidelberg. (Vol. 2, pp. 62–75).

23. Chinn, W. G., & Steenrod, N. E. (1966). *First concepts of topology: the geometry of mappings of segments, curves, circles, and disks*. MAA. (Vol. 18, pp. 74–78).

24. Chow, W. L. (1959). On the connectedness theorem in algebraic geometry. *American Journal of Mathematics*, *81*(4), 1033–1074.

25. Colonna, F. (2013). New criteria for boundedness and compactness of weighted composition operators mapping into the Bloch space. *Open Mathematics*, *11*(1), 55–73.

26. Comfort, W. W., & Hager, A. W. (1971). The projection mapping and other continuous functions on a product space. *Mathematica Scandinavica*, *28*(1), 77–90.

27. Conidis, C. (2012). Comparing theorems of hyperarithmetic analysis with the arithmetic Bolzano-Weierstrass theorem. *Transactions of the American Mathematical Society*, *364*(9), 4465–4494.

28. Conlon, L. (1964). The topology of certain spaces of paths on a compact symmetric space. *Transactions of the American Mathematical Society*, *112*(2), 228–248.

29. Davis, J. F., & Kirk, P. (2001). *Lecture notes in algebraic topology*. Providence: American Mathematical Society. (Vol. 35, pp. 367).

30. Einsiedler, M., & Ward, T. (2017). *Functional analysis, spectral theory, and applications*. Berlin: Springer. (Vol. 276, pp. 41–45).

31. Ekici, E., & Noiri, T. (2008). Connectedness in ideal topological spaces. *Novi Sad J. Math*, *38*(2), 65–70.

32. Erné, M. (2009). Infinite distributive laws versus local connectedness and compactness properties. *Topology and Its Applications*, *156*(12), 2054–2069.

33. Fischer, H., Repovš, D., Virk, Ž., & Zastrow, A. (2011). On semilocally simply connected spaces. *Topology and its Applications*, *158*(3), 397–408.

34. Fitch, F. B. (1950). A demonstrably consistent mathematics—part I. *The Journal of Symbolic Logic*, *15*(1), 17–24.

35. Font, J. J., & Sanchis, M. (2015). Sequentially compact subsets and monotone functions: an application to fuzzy theory. Topology and its Applications, 192, 113–122.

36. Gänssler, P. (1971). Compactness and sequential compactness in spaces of measures. Zeitschrift für Wahrscheinlichkeitstheorie und verwandte Gebiete, 17(2), 124–146.

37. Geba, K. (1964). Algebraic topology methods in the theory of compact fields in Banach spaces. Fund. Math, 54, 177–209.

38. Ghosh, A., Khan, M. A., & Uyanik, M. (2022). The intermediate value theorem and decision-making in psychology and economics: an expositional consolidation. *Games*, *13*(4), 51.

39. Ginsburg, J. (1975). Some results on the countable compactness and pseudocompactness of hyperspaces. *Canadian Journal of Mathematics*, *27*(6), 1392–1399.

40. Gronwall, T. H. (1919). Note on the derivatives with respect to a parameter of the solutions of a system of differential equations. *Annals of Mathematics*, *20*(4), 292–296.

41. Guo, T. (2008). The relation of Banach-Alaoglu theorem and Banach-Bourbaki-Kakutani-Šmulian theorem in complete random normed modules to stratification structure. *Science in China Series A: Mathematics*, *51*, 1651–1663.

42. Haydon, R. (1977). On Banach spaces which contain l 1 (τ) and types of measures on compact spaces. *Israel Journal of Mathematics*, *28*, 313–324.

43. Hofmann, K. H., & Morris, S. A. (2007). *The Lie theory of connected pro-Lie groups: a structure theory for pro-Lie algebras, pro-Lie groups, and connected locally compact groups.* European Mathematical Society. (Vol. 2, pp. 45–50).

44. Horn, W. (1970). Some fixed point theorems for compact maps and flows in Banach spaces. *Transactions of the American Mathematical Society*, *149*(2), 391–404.

45. Hsiao, G. C., & Wendland, W. L. (2008). *Boundary integral equations.* Springer Berlin Heidelberg. (Vol. 3, pp. 25–94).

46. Jäger, G. (1998). Compactness and connectedness as absolute properties in fuzzy topological spaces. *Fuzzy sets and Systems*, *94*(3), 405–410.

47. Janos, L., & Rosenfeld, A. (1983). Digital connectedness: An algebraic approach. *Pattern Recognition Letters*, *1*(3), 135–139.

48. Janssen, A., & Marohn, F. (1994). On statistical information of extreme order statistics, local extreme value alternatives, and Poisson point processes. *Journal of multivariate analysis*, *48*(1), 1–30.

49. Jayanti, S. V., & Tarjan, R. E. (2016, July). A randomized concurrent algorithm for disjoint set union. In *Proceedings of the 2016 ACM Symposium on Principles of Distributed Computing* (Vol. 3, pp. 75–82).

50. Jha, V., & Triar, U. S. (2021). A novel maximum power point tracking technique based on extreme value theorem for photovoltaic systems. *International Journal of Power Electronics*, *13*(3), 354–379.

51. Joseph, J. E., Kwack, M. H., & Nayar, B. M. (1999). Sequentially functionally compact and sequentially C-compact spaces. *Scientiae Mathematicae*, *2*(2), 187–194.

52. Juhász, I., & Szentmiklóssy, Z. (1993). Sequential compactness versus pseudo-radiality in compact spaces. *Topology and its Applications*, *50*(1), 47–53.

53. Keesling, J. (1974). Shape theory and compact connected abelian topological groups. *Transactions of the American Mathematical Society*, *194*, 349–358.

54. Keremedis, K. (2016). On sequentially compact and related notions of compactness of metric spaces in ZF. *Bull. Polish Acad. Sci. Math*, *64*, 29–46.

55. Klein, C. (1986). Arzela-Ascoli's theorem for Riemann-integrable functions on compact spaces. *manuscripta mathematica*, *55*(3), 403–418.

56. Kohli, J. K., & Das, A. K. (2006). A class of spaces containing all generalized absolutely closed (almost compact) spaces. Applied General Topology, 7(2), 233–244.

57. Koppelman, W. (1964). Spectral multiplicity theory for a class of singular integral operators. *Transactions of the American Mathematical Society*, *113*(1), 87–100.

58. Krukowski, M. (2018). Arzela-Ascoli theorem via the Wallman compactification. *Quaestiones Mathematicae*, *41*(3), 349–357.

59. Kumar, N. K. (2023). Equivalence among the Banach-Alaoglu theorem, Ascoli-Arzela theorem and Tychonoff's theorem. *Journal of the Institute of Engineering*, *17*(1), 8–11.

60. Lawrence, L. B. (1990). Paracompact product spaces defined by ultrafilters over the index set. *Proceedings of the American Mathematical Society*, *108*(2), 513–519.

61. Lianqiang, N., Xin, C., Min, P., & Gang, Z. (2017). Connected components labeling based on union-find operations applied to connected branches. *Journal of Intelligent & Fuzzy Systems*, *32*(5), 3739–3748.

62. Lima, A. (1978). Intersection properties of balls in spaces of compact operators. In *Annales de l'institut Fourier* (Vol. 28, pp. 35–65).

63. Liow, K. H., & Huang, Y. (2018). The dynamics of volatility connectedness in international real estate investment trusts. *Journal of International Financial Markets, Institutions and Money*, *55*, 195–210.

64. Macbeath, A. M. (1964). Groups of homeomorphisms of a simply connected space. *Annals of Mathematics*, *79*(3), 473–488.

65. Maciejewski, A. A., & Fox, J. J. (1993). Path planning and the topology of configuration space. *IEEE Transactions on Robotics and Automation*, *9*(4), 444–456.

66. Massey, W. S. (2019). A Basic Course in Algebraic Topology. Springer. (Vol. 127, pp. 76–80).

67. McCammon, S., & Hollinger, G. A. (2021). Topological path planning for autonomous information gathering. Autonomous Robots, 45(6), 821–842.

68. Mensi, W., Gubareva, M., Teplova, T., & Kang, S. H. (2023). Spillover and connectedness among G7 real estate investment trusts: The effects of investor sentiment and global factors. *The North American Journal of Economics and Finance*, *66*, 101919.

69. Michel, J. B., Thaunat, O., Houard, X., Meilhac, O., Caligiuri, G., & Nicoletti, A. (2007). Topological determinants and consequences of adventitial responses to arterial wall injury. *Arteriosclerosis, Thrombosis, and Vascular Biology*, *27*(6), 1259–1268.

70. Miteva, M. S., & Lazarova, L. K. (2018). Notion For Connectedness and Path Connectedness in Some Type of Topological Spaces. *Balkan Journal of Applied Mathematics and Informatics*, *1*(1), 55–62.

71. Monteiro, P. K. (1987). Some results on the existence of utility functions on path connected spaces. *Journal of Mathematical Economics*, *16*(2), 147–156.

72. Moore, R. L. (1919). On the most general class L of Fréchet in which the Heine-Borel-Lebesgue theorem holds true. *Proceedings of the National Academy of Sciences*, *5*(6), 206–210.

73. Mordukhovich, B. S., & Wang, B. (2001). Sequential normal compactness in variational analysis. *Nonlinear Analysis: Theory, Methods & Applications*, *47*(2), 717–728.

74. Motzkin, T. (1951). The lines and planes connecting the points of a finite set. *Transactions of the American Mathematical Society*, 451–464.

75. Neumann-Lara, V., & Wilson, R. G. (1986). Some properties of essentially connected and maximally connected spaces. *Houston J. Math*, *12*(3), 419–429.

76. Nguyen, T. T. H., Balli, F., Balli, H. O., & Syed, I. (2022). Direct real estate, securitized real estate, and equity market dynamic connectedness. *Applied Economics*, *54*(23), 2658–2677.

77. Nowak, M. (2011). Conditional weak compactness and weak sequential completeness in vector-valued function spaces. *Indagationes Mathematicae*, *21*(1–2), 40–51.

78. O'Meara, P. (1971). On paracompactness in function spaces with the compact-open topology. *Proceedings of the American Mathematical Society*, *29*(1), 183–189.

79. Pang, B., & Shi, F. G. (2014). Degrees of compactness in (L, M)-fuzzy convergence spaces and its applications. *Fuzzy Sets and Systems*, *251*, 1–22.

80. Paterson, A., & Welch, A. (2005). Tychonoff's theorem for locally compact spaces and an elementary approach to the topology of path spaces. *Proceedings of the American Mathematical Society*, *133*(9), 2761–2770.

81. Perfekt, R. (1997). Extreme value theory for a class of Markov chains with values in Rd. *Advances in Applied Probability*, *29*(1), 138–164.

82. Podczeck, K. (2008). On the convexity and compactness of the integral of a Banach space valued correspondence. *Journal of Mathematical Economics*, *44*(7–8), 836–852.

83. Rettinger, R., & Weihrauch, K. (2013). Products of effective topological spaces and a uniformly computable Tychonoff Theorem. *Logical Methods in Computer Science*, *9*, 85–88.

84. Russ, S. B. (1980). A translation of Bolzano's paper on the intermediate value theorem. *Historia Mathematica*, *7*(2), 156–185.

85. Sanders, S. (2021). Splittings and robustness for the Heine-Borel theorem. In *Connecting with Computability: 17th Conference on Computability in Europe, CiE 2021, Virtual Event, Ghent, July 5–9, 2021, Proceedings.* Springer International Publishing. *17* (Vol. 2, pp. 395–406).

86. Sell, G. R. (1967). Nonautonomous differential equations and topological dynamics I. The basic theory. *Transactions of the American Mathematical Society*, *127*(2), 241–262.

87. Serrano, D. H., Hernández-Serrano, J., & Gómez, D. S. (2020). Simplicial degree in complex networks. Applications of topological data analysis to network science. *Chaos, Solitons & Fractals*, *137*, 109839.

88. Sizemore, A. E., Phillips-Cremins, J. E., Ghrist, R., & Bassett, D. S. (2019). The importance of the whole: topological data analysis for the network neuroscientist. *Network Neuroscience*, *3*(3), 656–673.

89. So, S. S. (1991). Sequences, Mathematical Induction, and the Heine-Borel Theorem. *Missouri Journal of Mathematical Sciences*, *3*(2), 84–90.

90. Tanaka, Y. (1986). Decompositions of spaces determined by compact subsets. *Proceedings of the American Mathematical Society*, *97*(3), 549–555.

91. Treybig, L. B. (1964). Concerning continuous images of compact ordered spaces. *Proceedings of the American Mathematical Society*, *15*(6), 866–871.

92. Vaughan, J. E. (1984). Countably compact and sequentially compact spaces. In *Handbook of set-theoretic topology.* North-Holland. (Vol. 2, pp. 569–602).

93. Virk, Ž., & Zastrow, A. (2017). A new topology on the universal path space. *Topology and its Applications*, *231*, 186–196.

94. Wheeler, R. F. (1981). Weak and pointwise compactness in the space of bounded continuous functions. *Transactions of the American Mathematical Society*, *266*(2), 515–530.

95. Witten, E. (2016). Fermion path integrals and topological phases. *Reviews of Modern Physics*, *88*(3), 035001.

96. Zastrow, A. (1994). The second van-Kampen theorem for topological spaces. *Topology and its Applications*, *59*(3), 201–232.

97. Zou, H., Ahmed, M., Tariq, Q., & Khan, K. A. (2024). Time-varying connectedness between global economic policy uncertainty and regional real estate markets: evidence from TVP-VAR extended joint connectedness approach. *International Journal of Housing Markets and Analysis*, *17*(1), 79–95.

CHAPTER

4 Separation Axioms

LEARNING OBJECTIVES

After studying this chapter, you will be able to:

1. Understand the hierarchical structure of separation axioms in topological spaces, including T0, T1, and T2 spaces.
2. Differentiate between regular and normal spaces, and identify their significance in topology.
3. Apply Urysohn's Lemma to construct continuous functions between spaces and solve topological problems.
4. Utilize the Tietze Extension Theorem to extend continuous functions defined on subsets to the entire space.
5. Analyze the proof techniques behind Urysohn's Lemma and the Tietze Extension Theorem.
6. Explore practical applications of Urysohn's Lemma and the Tietze Extension Theorem in various topological contexts.

KEY TERMS FROM THIS CHAPTER

T0 Spaces

T1 Spaces

T2 Spaces (Hausdorff Spaces)

Regular Spaces

Normal Spaces

Urysohn's Lemma

Tietze Extension Theorem

Separation Axioms

UNIT INTRODUCTION

Separation Axioms in topology are foundational rules that categorize topological spaces according to their capacity to distinguish points and sets inside them. These axioms establish a systematic framework for comprehending the extent of "proximity" and "distance" within a space, which is crucial for examining its geometrical and topological characteristics. At the fundamental level, these axioms span from T0 to T4, with each subsequent axiom imposing more stringent conditions for separation. For example, the T0 axiom differentiates between points by utilizing open sets, guaranteeing that any pair of different points has a proximity that includes one point while excluding the other. Ascending the hierarchy, T1 spaces provide additional assurance that points are closed sets separately, while T2 (Hausdorff) spaces impose the most stringent separation requirement, ensuring that any two different points have non-overlapping open neighborhoods (Al-Shami, 2021) (Figure 4.1).

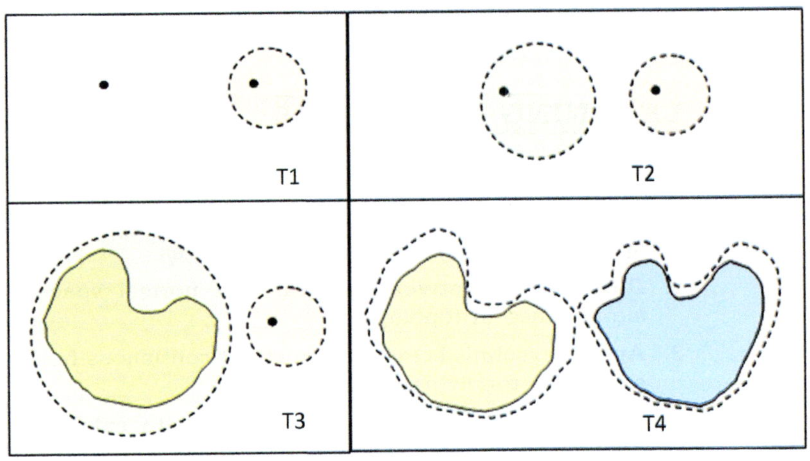

Figure 4.1. T1, T2, T3 and T4 Spaces in Separation Axioms in topology (Source: Math Strek, Creative Commons License)

The significance of the above separation axioms rests in their wide-ranging applicability in diverse areas of mathematics and more. Topology employs precise descriptions and categorization of spaces to enhance the understanding and development of fundamental theoretical concepts like compactness, connection, and continuity. Practical applications are found in areas such as analysis, where theorems that are crucial for function extension as well as space embedding rely on features obtained from separation axioms. Furthermore, in disciplines such as computational science and physical science (Ávila & Antonio, 2005), where the representation of intricate systems frequently requires consideration of spatial connections, these axioms serve as a meticulous foundation for developing programs and simulations that faithfully depict real-life situations. The study and use of separation axioms have a profound impact on mathematical theory and practical advancements in several fields. This highlights their lasting importance in contemporary mathematics and beyond (Jenavee et al., 2022).

4.1. HIERARCHIES OF SEPARATION

The following is an elucidation of the order of separation.

4.1.1. T0 Spaces

A "T0 space," commonly referred to as a Kolmogorov space, is a specific sort of topological space that meets a basic need for separation. In a T0 topological space, it has at least one open set which includes one of any two distinct points instead of the other (Dritschel & Rovnyak, 1996). Formally, a topological space (X) is T0 if, for every pair of distinct points $(x, y \in X)$, there exists an open set $(U \subset X)$ such that either $(x \in U)$ and $(y \notin U)$ or $(y \in U)$ and $(x \notin U)$.

Remember

Remember the hierarchy of separation axioms in topology, from T0 spaces to normal spaces, which provide increasingly stringent conditions on how points and closed sets can be separated.

4.1.1.1. Illustrative Examples and Counter Examples

To gain a deeper comprehension of T0 spaces, it is beneficial to examine the following illustrations:

1. Example 1: The Real Line (R): The real line equipped with the conventional topology, whereby open sets are defined as intervals, is considered a T0 space. For every two different points (x) and (y) on the real line, there always exists an open interval that contains one point but not the other (Lassez & Marriott, 1987).
2. Example 2: The Sierpiński Space: This space consists of two points, say (a) and (b), with the topology $(\{\varnothing, \{a\}, \{a,b\}\})$. This is a T0 space because there is an open set $(\{a\})$ that contains (a) but not (b).
3. Counterexample: The Indiscrete Topology is a case that disproves a certain claim or proposition. Let (X) be a set with more than one member, and let it be equipped with

the indiscrete topology, which means that only the empty set and the full set are considered open (Clarke et al., 2000). This space cannot be classified as a T0 space due to the absence of open sets that can differentiate between any two separate points in (X).

4.1.1.2. Role in Topological Structures

Topologically, T0 spaces are the fundamental units in the order of separation axioms. They serve as the least strong form of separation, guaranteeing that separate points can be differentiated by open sets in terms of topology. This characteristic is essential in numerous domains of topology and analytical mathematics for multiple reasons:

1. Uniqueness of Limits: In a T0 space, the limits of sequences or nets, should they exist, are unique. This distinctiveness is essential for guaranteeing specified topological configurations (Estrada & Uriarte, 2001).
2. T0 spaces serve as the foundation for establishing stricter separation axioms, such as T1 (Frechet) and T2 (Hausdorff) spaces. All of the higher axioms extend the separation qualities of T0 spaces by incorporating additional criteria to provide a greater level of distinction between points.
3. Implementations in Data Analysis: T0 spaces are valuable in practical fields such as analysis of data and computer science, as they aid in the creation and examination of spaces where the differentiation of points based on specific qualities is crucial (Chedin & Benham, 2020).

Zero-dimensional spaces, despite their low separation requirements, have a substantial impact on the overall understanding of topological spaces. They serve as the initial stage towards the exploration of more intricate and well-organized topological characteristics.

4.1.2. T1 Spaces

A "T1 space," also referred to as a Frechet space, is a specific sort of topological space in which every pair of different points has separate neighborhoods that do not overlap.

Formally, a topological space (X) is T1 if, for every pair of distinct points $(x, y \in X)$, there exist open sets (U) and (V) such that $(x \in U),(y \notin U)$, and $(y \in V),(x \notin V)$. This condition ensures that points can be "separated" by open sets in both directions (Soltanian & Peck, 2001).

4.1.2.1. Comparative Analysis with T0 Spaces

T1 spaces are a logical continuation of T0 spaces. In T0 spaces, it is necessary for there to be at least one open set that can differentiate between any two different points. However, in T1 spaces, each point must have an open set that does not include the other point. As a result, every T1 space is also a T0 space, but not every T0 space is a T1 space. In T1 spaces, the presence of an additional separation requirement results in more intricate topological structures, hence improving the capacity to differentiate between points (Goldfarb et al., 2004).

4.1.2.2. Practical Examples and Common Misconceptions

Example 1: Discrete Topology: A set (X) equipped with the discrete topology, in which every subset is open, is classified as a T1 space. Each point is considered to be isolated, and individual sets ({x}) and ({y}) are used as open sets to distinguish any different points (x) and (y).

1. Widespread Misunderstanding: There is a prevalent misconception that T1 spaces require a basis of open sets that only contain individual points. In a T1 space, every singleton set is closed, but it is not necessary for the space to have a basis consisting only of such sets (Coley & Tanner, 2012).
2. Significant characteristics and essential principles: Singletons that are not open or accessible: In T1 spaces, each set consisting of a single element, denoted as ({x}), is considered to be a closed set. This conclusion is a straightforward consequence of the concept, as the set ({x}) may be isolated from all other points via open sets.
3. Finite sets exhibit closure: In a T1 space, every finite subset is considered closed. This concept expands on the idea of closed singletons by stating that the finite combination of closed sets is also closed (Ling et al., 2016).
4. Convergence in T1 spaces guarantees that sequences, nets, and filters maintain distinct bounds when they converge. The concept of uniqueness is crucial in the fields of analysis and topology when it comes to establishing clearly defined bounds.T1 implies T0. A T1 space is a subset of a T0 space, as the T1 criterion is more stringent than the T0 condition.

T1 spaces improve the separation qualities of T0 spaces by guaranteeing that neighborhoods around different locations do not overlap. T1 spaces play a vital role in the hierarchy of separation axioms as they offer a more sophisticated framework for topological analysis (Gomez, 2008).

4.1.3. T2 Spaces (Hausdorff)

A "T2 space," also known as a "Hausdorff space," is a type of topological space in which it is possible to find two different locations that may be completely isolated from each other by non-overlapping neighborhoods. Formally, a topological space (X) is T2 if, for every pair of distinct points $(x, y \in X)$, there exist disjoint open sets (U) and (V) such that $(x \in U)$ and $(y \in V)$. This condition guarantees that each pair of distinct points is "pushed apart" by the topology, ensuring a high level of separation (Huttenlocher et al., 1993) (Figure 4.2).

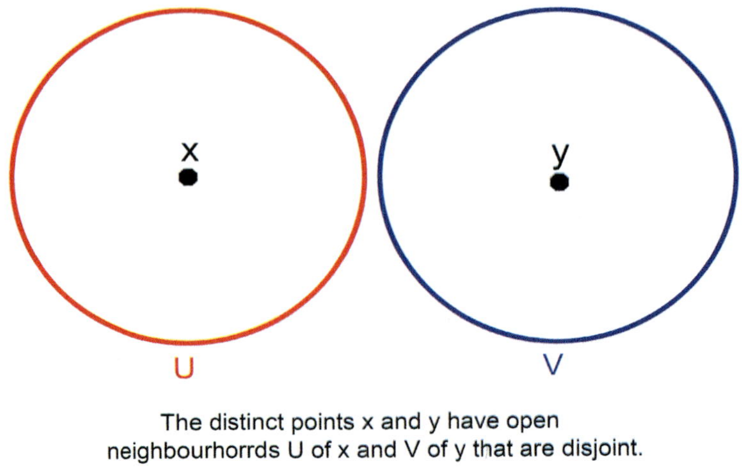

Figure 4.2. T2 Hausdorff Topological Spaces (Source: Math Online, Creative Commons License)

4.1.3.1. Relationship to T0 and T1 Spaces

T2 spaces possess greater strength compared to both T0 and T1 areas. Every T2 space is also T1 since if two points can be isolated by non-overlapping neighborhoods, they can definitely be isolated by neighborhoods that do not include the other point. Furthermore, T2 spaces are T0 because every pair of different points may be differentiated by not less than one open set that contains one point and excludes the other. Consequently, the hierarchy is as follows: every T2 space is also a T1 space, and every T1 space is also a T0 space (Kopperman & Tkachuk, 2003).

4.1.3.2. Representative Examples

1. Example 1: The set of real numbers, denoted by (R): The real line, equipped with the usual topology, is a well-known example of a T2 space. For any two distinct points (x) and (y), open intervals $((x-\epsilon, x+\epsilon))$ and $((y-\epsilon, y+\epsilon))$ with sufficiently small (ϵ) will be disjoint and contain (x) and (y), respectively.
2. Example 2: Euclidean Space (R^n): Likewise, any Euclidean space equipped with its typical topology is a T2 space. Discrete open balls can be located around any two separate points (Butler et al., 1983).
3. Non-Example: The Indiscrete Topology: A set (X) equipped with the indiscrete topology, which only considers the empty set and (X) itself as open, does not satisfy the T2 separation axiom. This is because it is impossible to find two different points in (X) that can be separated by open sets that have no points in common (Skånér et al., 2005).

4.1.3.3. Core Properties and Theorems

1. The uniqueness of limit points is a notable characteristic of T2 spaces, as it guarantees that limits of sequences, nets, or filters have only one possible value. If a sequence $((x^n))$ in a T2 space converges to both (x) and (y), then (x) and (y) must be equal. This distinctiveness is essential in analysis and guarantees that boundaries are clearly established.
2. Convergence and Separation: Convergence of sequences and separation of points are intricately interconnected in T2 spaces. For example, when a sequence converges, the neighborhood system around the limit point must get more precise as the sequence advances, in accordance with the separation of points.
3. Continuous functions in T2 spaces exhibit very regular features. For instance, if a continuous function is applied to a compact set in a Hausdorff space, the resulting picture will also be compact. This trait is crucial in the field of functional analysis and has numerous applications.
4. The product of two T2 spaces, when equipped with the product topology, is also a T2 space. The property of being closed under products is advantageous for creating new Hausdorff spaces from preexisting ones (Karaisa & Başar, 2014).

4.1.3.4. Significance in Analysis and Broader Topology

T2 spaces, often known as Hausdorff spaces, have a significant impact on applied as well as pure mathematics. Their distinct qualities make them crucial in analysis, wherein the uniqueness of limits and clearly defined convergence are vital. The Hausdorff condition is implicitly relied upon by several key theorems, including the Intermediate Value Theorem as well as the Extreme Value Theorem (Shavitt & Weinsberg, 2011).

Hausdorff spaces in topology offer a sturdy foundation for analyzing continuous functions, convergence, and compactness. Hausdorff properties are fundamental for various advanced notions, including metric spaces and topological manifolds. Furthermore, in the contemporary study, T2 spaces play a crucial role in disciplines such as differential algebra, polynomial topology, and functional analysis, highlighting their wide-ranging relevance and significance.

To put it concisely, T2 spaces guarantee a significant level of distinction between points, which is crucial for a topological structure to function properly. The features and theorems of these concepts are fundamental in both analysis and topology, underscoring their importance in the wider mathematical field (Siddiqui et al., 2018).

4.2. ADVANCED SEPARATION PROPERTIES

Following is the explanation of advanced separation properties.

4.2.1. Regular Spaces: Precise Local Separation

A "regular space" refers to a topological space in which points can be distinguished from closed sets using neighborhoods. Formally, a topological space (X) is regular if for every point $(x \in X)$ and every closed set $(C \subset X)$ not containing (x), there exist disjoint open sets (U) and (V) such that $(x \in U)$ and $(C \subset V)$ (Figure 4.3).

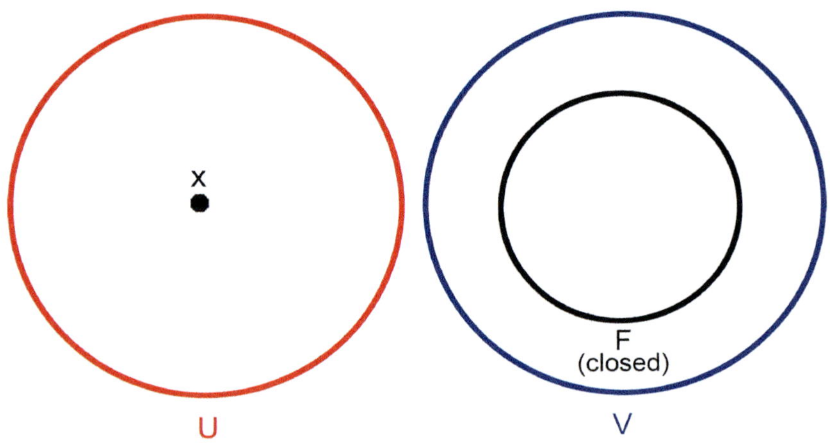

Figure 4.3. T3 Regular Hausdorff Topological Spaces (Source: Math Online, Creative Commons License)

This trait guarantees a precise level of segregation, where not only individual points can be distinguished, but points can also be deliberately isolated from closed sets (Auscher & Hytönen, 2013) (Figure 4.4).

Did you know?

Did you know that regular spaces generalize T2 (Hausdorff) spaces by ensuring that points and closed sets can be separated by disjoint open sets, extending the separation properties beyond the Hausdorff condition?

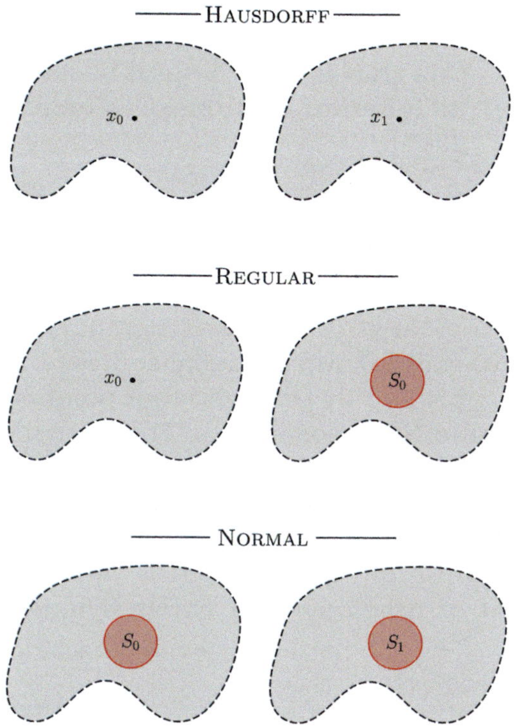

Figure 4.4. An illustration of some of the separation axioms. Grey amorphous broken-outline regions indicate open sets surrounding disjoint closed sets or points: red solid-outline circles denote closed sets while black dots represent points (Source: Osmium, Creative Commons License)

4.2.1.1. Interaction with T0, T1, and T2 Spaces

Regular spaces have intriguing intersections with T0, T1, and T2 spaces. A T1 space that satisfies the Hausdorff separation axiom is commonly referred to as a "T3 space" or a "regular Hausdorff space." In a T3 space, points are not only separated by neighbors (like in T1 spaces) but they are also isolated from closed sets, hence increasing the separation qualities to a greater extent. Regularity is frequently assessed in relation to T1 or T2 qualities to guarantee the resilience of topological structures (Ludwig & Affleck, 1994).

4.2.1.2. Notable Examples and Exceptions

Example 1: In the context of metric spaces, it can be stated that any metric space is periodic. Given a point (x) and a closed set (C) in a metric space, it is possible to utilize the metric to create separate open balls about (x) and (C) that do not intersect (Black & Smith, 1993).

The Sierpi⊠ski Space refers to a mathematical concept that is named after the Polish mathematician Wacław Sierpi⊠ski. It is a topological space that consists of three points and is commonly used in the study. The Sierpi⊠ski space exhibits the property of being a T0 space, but, it does not possess the property of being regular. The topology lacks the ability to separate points from closed subsets in the manner necessary for regular spaces (Landry, 2024).

4.2.1.3. Key Theorems and Applications

1. The application of Urysohn's Lemma is limited to normal spaces, which are topological spaces that satisfy both the regularity and T1 axioms. Nevertheless, comprehending the principles that underlie Urysohn's Lemma is crucial for grasping the nature of regular spaces. The lemma asserts that in a normal space if there are two separate closed sets, there will always be a continuous function that can transfer these sets to different values. This outcome is crucial in the construction of continuous functions with defined attributes and highlights the significance of separation axioms (Radulescu & Radulescu, 1980).

2. The importance of regular spaces is significant in the examination of function spaces. For example, in a typical space, continuous functions can be extended in a manner that maintains regularity. This trait is valuable in the fields of analysis and functional analysis, as it allows for the extension of functions while preserving their topological qualities (Fekete et al., 2019).

4.2.1.4. Core Theorems and Applications

1. Characterization of Regularity as a Theoretical Concept: For every point (x) and every neighborhood (U) of (x), there exists a neighborhood V(x) such that the closure of (V) is contained inside (U). This is the only condition under which a space (X) is considered regular. With this characterization, a realistic criterion for determining whether or not topological spaces are regular can be accomplished (Kutbi et al., 2015).
2. Regularity and Product Spaces; A Theoretical Perspective: This does not necessarily mean that the result of two regular areas is also regular. This finding sheds light on the complex nature of regularity and emphasizes the importance of taking into account the specific factors that make it possible for product spaces to maintain regularity.
3. Use in practice: Continuous functions can be created with greater ease in regular spaces because of the separation of points and closed sets that occur in these spaces. When it comes to function spaces, where it is frequently necessary to design functions with particular qualities, this trait is very helpful because of its inherent usefulness (Mursaleen, 2004).

When it comes to the hierarchy of separation qualities in topology, regular spaces are an essential step that must be taken. Regular spaces offer an advanced form of topological structure that is important in both conceptual and practical contexts. This is accomplished by ensuring that points may be isolated from closed sets. The connection with T0, T1, and T2 spaces, in addition to important theorems and applications like Urysohn's Lemma, helps to highlight the significance of regular spaces within the context of topology as a whole (Chaperon, 2008).

4.2.2. Normal Spaces: Higher-Order Separation

One definition of a "normal space" describes it as a topological space in which any two disjoint close sets can be divided by disjoint residential areas. Formally, a topological space (X) is normal if for every pair of disjoint closed sets $(A, B \subset X)$, there exist disjoint open sets (U) and (V) such that $(A \subset U)$ and $(B \subset V)$. It is only necessary for points to have been separated from closed sets in order to expand the idea of separation in standard spaces, which is this property's extension (Yamashita, 1992) (Figure 4.5).

Figure 4.5. T4 Normal Hausdorff Topological Spaces (Source: Math Online, Creative Commons License)

4.2.2.1. Comparative Analysis with Regular and T2 Spaces

The features of separation that are possessed by regular spaces and T2 spaces are utilized by normal spaces. A "T4 space" (sometimes referred to as a "normal Hausdorff space") is the name given to a normal T1 space. This means that every T4 space is a T3, and every T3 space is a T2. To summarize this hierarchy, the following is what can be said:

i. T2 (Hausdorff) Spaces: Any two different points can be separated by neighborhoods that are not connected.
ii. When it comes to T3 (Regular Hausdorff) Spaces, every point and a closed set that does not contain it can be separated by neighbors that are not connected.
iii. T4 (Normal Hausdorff) Spaces: any pair of disjoint closed sets may be divided by disjoint neighborhoods. This pertains to all disjoint closed sets (Carolan, 2017).

In their most basic form, normal spaces are regular spaces that have an additional layer of separation. This gives them a great degree of structure and makes them helpful in a variety of topological and analytical fields (Rodriguez et al., 2018).

4.2.2.2. Important Examples and Counterexamples

An illustration of the Euclidean space (R^n): By the same token, any Euclidean space that possesses the standard topology is considered normal. In order to extend the separation properties of the real line to higher dimensions, it is possible to separate disjoint closed sets using open balls (Kernighan & Plauger, 1974).

An alternative illustration is the Sorgenfrey Plane: One example of a space that does not conform to the category of normal spaces is the Sorgenfrey plane, which is defined as (R^2) with the Sorgenfrey topology. Within the confines of this topology, it is feasible to create two disjoint closed sets that are incapable of being separated by disjoint open sets (Neiskashova & Fomina, 2020).

4.2.2.3. Fundamental Theorems and Applications

1. A Practical Application of the Tietze Extension Theorem: The Tietze Extension Theorem is one of the most important results for normal spaces. Under the assumption that (X) is a regular space and (A) is a closed subset of (X), it is stated that any continuous function (f:A→R) can be expanded to a continuous function (F:X→R). In the fields of analysis and topology, this theorem is extremely useful since it enables the generalization of continuous functions that are defined on closed sets to the complete space without compromising the continuity of the function (Lancaster, 1977).
2. In addition to being a fundamental result that may be applied to normal spaces, Urysohn's Lemma is also a fundamental result. It asserts that there is a continuous function $\left(f : X \rightarrow [0,1]\right)$ such that $\left(f(A) = \{0\}\right)$ and $\left(f(B) \quad \{1\}\right)$ that can be implemented for any two disjoint closed sets (A) and (B) that are located in a normal space (X). The construction of continuous functions that separate closed sets is a crucial step in a wide variety of proofs and applications, and this lemma is vital for its construction (Sylvester, 1878).
3. The great amount of separation that normal spaces possess makes them an important component in both topology and analysis. This is because normal spaces play a vital role in both of these areas. The study of compactifications, metrization theorems, and the development of continuous functions are all aspects of mathematics that make use of them. An example of this would be the fact that numerous metrization theorems assert that, under specific circumstances, a normal space can be assigned a metric that results in the same topology. In analysis, where working with metric spaces frequently simplifies issues and procedures, this is an extremely important point to keep in mind (Faure, 2002).

Within the hierarchy of topological spaces, normal spaces are a higher-order separation attribute. In order to offer a stable framework for the construction of continuous functions and the extension of such functions, normal spaces ensure that any two disjoint closed sets can be separated by disjoint neighborhoods. This allows for the development of continuous functions. The Tietze Extension Theorem and Urysohn's Lemma are two important results that bring attention to the significance of normal spaces in both theoretical and practical settings. As a result of these features and theorems, normal spaces are an essential idea in the field of topology, as well as in the applications of topology to analysis and other branches of mathematics (Luchko, 2020).

4.3. URYSOHN'S LEMMA: CONSTRUCTING CONTINUOUS FUNCTIONS

4.3.1. Formal Statement of Urysohn's Lemma

Urysohn's Lemma is a result that is considered to be a foundational finding in topology, particularly in relation to normal spaces.

Let (X) be a normal topological space, and let (A) and (B) be two disjoint closed subsets of (X). Then there exists a continuous function $(f:X\to[0,1])$ such that $(f(a)=0)$ for all $(a\in A)$ and $(f(b)=1)$ for all $(b\in B)$ (Kreinovich, 2010).

Urysohn's Lemma ensures the presence of a continuous function that assigns specific values (0 and 1) to two separate closed sets and changes continuously within the interval ([0, 1]) over the entire space (Morsi, 1991).

4.3.1.1. Preconditions and Necessary Context

In order to apply Urysohn's Lemma, certain prerequisites must be fulfilled:

1. Requirement: The space (X) must satisfy the condition of being normal. This implies that it is possible to find two independent neighborhoods that do not overlap, such that each neighborhood contains one of the disjoint closed sets in (X). Normality is an essential condition because the lemma relies on the capability of continuous functions to distinguish closed sets (Cigler, 1999).
2. Disjoint closed sets: The sets (A) and (B) must be both closed and disjoint in (X). This lack of overlap is essential for establishing the function (f) that assigns distinct values (0 and 1) to (A) and (B), respectively.
3. Values of a mathematical function: The function (f) assigns elements from the space (X) to the closed interval ([0,1]). The points in set (A) are assigned the value 0, while the points in set (B) are allocated the value 1. This assignment ensures that the function (f) respects the separation of these sets (Connor, 1993).

4.3.1.2. Example for Context

Consider the real line (R) with the standard topology. The sets $(A=\{x\leq 0\})$ and $(B=\{x\geq 1\})$ are closed and disjoint. According to Urysohn's Lemma, there exists a continuous function $(f: \mathbb{R} \to [0,1])$ such that $(f(x)=0)$ for all $(x\leq 0)$ and $(f(x)=1)$ for all $(x\geq 1)$.

$$f(x)=0 \;\&\text{“}\{if\}x\leq 0$$

$$x \;\&\text{“}\{if\}0<x<1$$

$$1 \;\&\text{“}\{if\}x\geq 1$$

This example demonstrates the fundamental concept of Urysohn's Lemma: constructing a continuous function that connects two separate closed sets in a normal space (Holmes, 2006).

Tip

Urysohn's Lemma is essential in topology for constructing continuous functions that separate points from closed sets in normal spaces, demonstrating the powerful use of topology in function construction.

4.3.2. Detailed Proof of Urysohn's Lemma

To prove Urysohn's Lemma, construct a continuous function $(f:X\rightarrow[0,1])$ step by step. To achieve this objective, establish a continuous function that assigns the value 0 to set (A) and the value 1 to set (B), where (X) is a typical topological space and (A) and (B) are two separate closed sets (Pedersen, 1997).

Step 1: Constructing a Sequence of Open Sets

1. For each rational number $(r\in(0,1))$, construct open sets ( ) such that: $\left[A\subset U_r\subset \underline{U_r}\subset V_r\subset V_r\subset \underline{V_r}\subset X\setminus B\right]$
2. This construction relies on the normality of (X). For each (r), use the normality of (X) to separate (A) and (B) by open sets. Specifically, for each $(r\in(0,1))$, find disjoint open sets ( ) and (V_r) such that: $\left[A\subset U_r\subset \underline{U_r}\subset V_r\subset V_r\subset \underline{V_r}\subset X\setminus B\right]$

Step 2: Defining the Function (f)

1. Define $(f : X \to [0,1])$ using the sequence of open sets $(\{U_r\}).Set : [f(x) = supsup\{r \in Q \cap [0,1] : x \in U_r\}]$
2. By construction, $(f(x))$ will assign values between 0 and 1 to each $(x \in X)$, with $(f(x) = 0)$ for $(x \in A)and(f(x) = 1)for(x \in B)$.

Step 3: Verifying Continuity

1. To prove (f) is continuous, let \$(c \in [0,1])$ and consider the set $(f^{-1}([0,c)))$. This set can be expressed as: $[f^{-1}([0,c)) = \square_{r<c,r \in Q} U_r]$
2. Since (U_r) are open, and the union of open sets is also open. Hence, (f^{-1}([0, c))) is open in (X).
3. Similarly, consider $(f^{-1}((c,1])) : [f^{-1}((c,1]) = \square_{r>c,r \in Q} (X \setminus \underline{U_r})]$
4. Since $(X \setminus \underline{U_r})$ are open, the intersection of open sets is also open. Hence, $(f^{-1}((c,1]))$ is open in (X) (Tachtsis, 2019).

4.3.2.1. Discussion of Key Ideas and Techniques

The concept of normalcy is essential in the construction of the disjoint open sets (U^r) and (V^r) for every rational (r). The separation mentioned serves as the fundamental basis for the function (f) by guaranteeing that (f) can include all values inside the interval ([0,1]) while maintaining continuity (DeLozier & Rhodes, 2017).

Supremum Function Definition: The function $(f(x) = supsup\{r \in Q \cap [0,1] : x \in U_r\})$ leverages the density of rationals in $([0,1])$ to create a continuous mapping from (X) to $([0,1])$. The use of this supremum construction guarantees that function (f) preserves the properties of the open sets (U_r) and (V_r), while also seamlessly transitioning between the values of 0 and 1.

Continuity Verification: Demonstrating the continuity of (f) by analyzing the preimages of open sets in ([0,1]) showcases the resilience of the function. The requirement of having open set preimages ensures that the function (f) is continuous over the set (X).

Urysohn's Lemma is a valuable technique for creating continuous functions that can separate closed sets that do not overlap in normal spaces. The comprehensive proof emphasizes the significance of normalcy, the formation of open sets, and the notion of supremum, showcasing the sophistication and usefulness of this basic finding in topology (Nyumba et al., 2018).

4.3.3. Practical Applications of Urysohn's Lemma

4.3.3.1. Usage in Constructing Continuous Functions

Urysohn's Lemma is a potent instrument for generating continuous functions that distinguish closed sets in a normal space. This feature is essential in diverse domains of topology and research. Urysohn's Lemma offers a technique to construct a continuous function that can differentiate between two separate closed sets. This structure is crucial in demonstrations and practical applications that necessitate the continuous separation of sets, allowing for additional examination and modification of topological spaces (Kaplansky, 1947) (Figure 4.6).

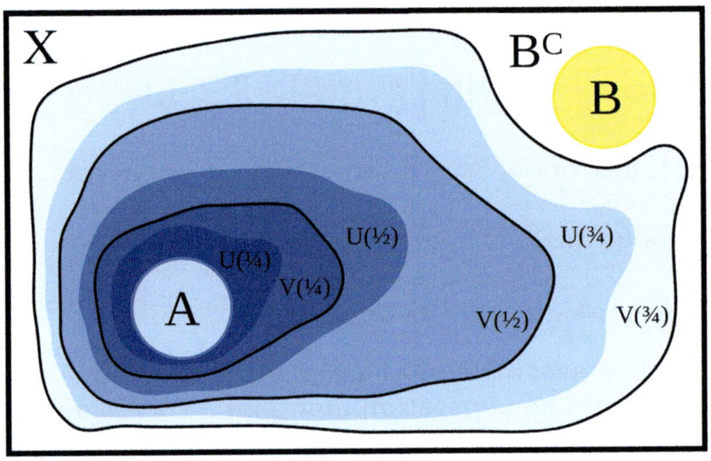

Figure 4.6. Illustration of the first few sets built as part of the proof (Source: Roman Salinas, Creative Commons License)

4.3.3.2. Application to Metrization Theorems

Urysohn's Lemma is commonly utilized in the context of Metrization theorems in order to demonstrate that particular topological spaces can be fitted with a metric that generates the same topology. According to the Urysohn Metrization Theorem, a topological space can be provided with a metric if and only if it satisfies the conditions of being regular, second-countable, and Hausdorff (Bregman et al., 1984). This is the only condition that can a topological space be endowed with a metric. Urysohn's Lemma makes it easier to generate the continuous functions that are necessary to demonstrate that a space can be fitted with a metric that is consistent with their space. Considering the connection between normality and metrizability, it is clear that the lemma is of critical importance in both the understanding of metric spaces in topology and the advancement of these spaces (Miñana et al., 2023).

4.3.3.3. Role in Regular and Normal Spaces

When distinguishing between the many types of topological spaces, particularly regular and normal spaces, Urysohn's Lemma is an extremely important concept to understand. Through the use of the lemma, it is possible to generate continuous functions that are capable of distinguishing points from closed sets in topological spaces that are typical. Additionally, in a typical space, it is necessary to divide any two distinct closed sets in order to complete the process (Di Donato et al., 2022). The property of separation is of the utmost relevance for a wide variety of theoretical breakthroughs as well as practical applications. Individuals are able to get a more comprehensive understanding of the arrangement and properties of regular and normal spaces through the utilization of Urysohn's Lemma, which paves the way for more significant discoveries and practical applications in the field of topology (Heldermann, 1980).

4.3.3.4. Diverse Examples in Topological Contexts

1. Constructing Continuous Extensions: Urysohn's Lemma can be utilized in real-world scenarios to extend continuous functions that are characterized by a closed subset of a normal space to the complete space. This is accomplished by constructing continuous extensions. Due to the fact that it enables the seamless extension of functions that are specified on specific regions to wider domains, this is especially advantageous in the field of analysis and computational modeling (Kallio, 2018).
2. Compactifications: Urysohn's Lemma is applied in the process of compactifying a space, which involves embedding a space that is not compact into a space that is compact. This approach is described in more detail below. It is possible to develop compactifications that preserve the characteristics of the original topology if continuous functions that are able to differentiate between sets are created.
3. One of the applications of partitions of unity in the study of differential topology is the generation of smooth functions on manifolds that have specific characteristics. Because it provides a method for designing continuous functions that are accurately suited to the local properties of the manifold, Urysohn's Lemma makes it easier to create these functions (Fridy & Orhan, 1997).
4. Demonstrating Theorems in Analysis: A great number of theorems in real and functional analysis are dependent on the ability to generate continuous functions that have specific characteristics. Urysohn's Lemma is a key technique that enables precise manipulation of function behavior. It makes it easier to construct proofs and applications that require precision control (Lin et al., 2023).

Urysohn's Lemma is a key idea in topology that has extensive practical applications in the production of continuous functions, the testing of metrization theorems, and the grasp of the features of regular and normal spaces. It is recognized as one of the most important developments in the field of topology. One of the many practical applications of topology is the extension of functions, the compactification of spaces, and the formation of partitions of unity (Egenhofer & Herring, 1990).

4.4. TIETZE EXTENSION THEOREM: EXTENDING CONTINUOUS FUNCTIONS

In the field of topology, the "Tietze Extension Theorem" is a significant finding that deals with the extension of continuous functions that are defined on closed subsets of a topological space to the entire space (Grabiner, 1986).

Statement: Let $\left(X\right)$ be a normal topological space and let $\left(A\right)$ be a closed subset of $\left(X\right)$. Suppose $\left(f:A\rightarrow R\right)$ (or $\left(\tilde{o}\,o\{C\}\right)$) is a continuous function. Then there exists a continuous function $\left(F:X\rightarrow R\right)$ (or $\left(\tilde{o}\,o\{C\}\right)$) such that $\left(F|_{A}=f\right)$, where $\left(F|_{A}\right)$ denotes the restriction of $\left(F\right)$ to $\left(A\right)$.

Essential prerequisites and background information:

To apply the Tietze Extension Theorem, it is necessary for the following requirements to be met:

1. Space (X) is said to have normality. This implies that it is possible to find two independent open neighborhoods that do not overlap, such that each neighborhood contains one of the disjoint closed subsets of (X). Normality is essential because it enables the creation of continuous functions that can separate closed sets, a characteristic that is utilized in the demonstration of the theorem. A closed subset (A) is a subset of (X) that satisfies the property of being closed. Closedness guarantees that set (A) possesses a clearly defined border within the space (X), which simplifies the process of extending it (Pąk, 2014).
2. A continuous function (f) is defined on set (A) and maps to the set of real numbers ((R)) or complex numbers ((C)). The continuity of function (f) on set (A) is crucial as it serves as the initial step in constructing the extension (F).

Explanation:

The Tietze Extension Theorem guarantees the existence of a continuous extension $\left(F:X\rightarrow R\right)\left(or\left(C\right)\right)of\left(f\right)$ from $\left(A\right)$ to the entire space $\left(X\right)$. This extension $\left(F\right)$ preserves the continuity of $\left(f\right)$ on $\left(A\right)$, meaning $\left(F|_{A}=f\right)$.

Example:

Consider $(X=[0,1])$ with the standard topology and $(A=\{0\}\cup\{n\in N\})$, which is a closed subset of (X). Define $(f:A\to R)$ by $(f(x)=0)$ for $(x\in A)$. According to the Tietze Extension Theorem, there exists a continuous function $(F:[0,1]\to R)$ such that $(F(x)=0)\,for(x\in A)$ and $(F(x)=0)$ for $(x\in X\setminus A)$.

This theorem is not only significant from a theoretical standpoint, but it also has practical applications in a variety of areas of mathematics, including functional analysis, algebraic topology, and mathematical physics, among others. It enables mathematicians to expand their analysis from closed subsets to encompass complete spaces, which in turn makes it easier for them to get significant insights and apply their findings in a wider range of contexts when studying topological spaces (Kubiak, 1987).

4.4.1. Proof of Tietze Extension Theorem

To establish the validity of the Tietze Extension Theorem, systematically create a continuous extension (F) for a specified continuous function, given that (A) is a closed subset of a normal topological space (X).

Step 1: Constructing Local Continuous Extensions

Local Construction: Begin by constructing continuous functions $(g_n:X\to[0,1])$ for each $(n\in N)$, such that:

$g_n(x)$

$0, \&$ "$\{if\}x\in A$

$1, \&$ "$\{if\}x\in X\,Un$

where (U_n) is an open set containing (A) in (X).

Applying Urysohn's Lemma: Employ Urysohn's Lemma to guarantee the existence of functions (g_n). By the property of normality of (X), there exists a continuous function (g_n) for each (n) such that (g_n) is equal to 0 on (A) and equal to 1 outside a certain open neighborhood (U_n).

Step 2: Combining Local Extensions

Combining Functions: Define $\left(h_n(x)=\sum_{k=0}^{n}\frac{1}{2^k}g_k(x)\right)$. Each (h_n) is continuous and converges uniformly to a continuous function $(h:X\to[0,1])$.

The uniform convergence of (h_n) to (h) guarantees that (h) is continuous on (X) because the limit of a sequence of continuous functions is also continuous (Ciesielska & Ciesielski, 2017).

Step 3: Scaling and Modification

Scaling and Modification: Scale (h) to map into the range of (f). Let $(F(x) = f(x)h(x))$.

The continuity of (F) on (X) may be established by considering the continuity of (f) on (A) and (h) on (X). Since both (f) and (h) are continuous functions, their product (F) is also continuous on (X).

Example 1: Closed Interval

Consider $(X = [0,1])$ with $\left(A = \{0\} \cup \left\{\frac{1}{n} \mid n \in N\right\}\right)$, a closed subset of (X). Define $(f : A \to R) by (f(x) = 0)$ for $(x \in A)$.

By applying the Tietze Extension Theorem, we construct $(F : [0,1] \to R)$ such that $(F(x) = 0)$ for $(x \in A)$ and $(F(x) = 0)$ for $(x \in X \setminus A)$.

Examination of Key Methods and Ideas:

1. Regional Additions: By employing local continuous functions (g_n), the extension (F) preserves the characteristics of (f) on (A) while ensuring continuity.
2. Uniform convergence refers to the property of a sequence (h_n) that is formed from (g_n) and converges uniformly to (h). This property allows for a smooth transition in defining (F) over (X).
3. The presence of normality in (X) guarantees that (A) and its complement may be distinguished by open neighborhoods, which simplifies the creation of (g_n) by the utilization of Urysohn's Lemma.

The Tietze Extension Theorem offers a reliable approach to extending continuous functions that are defined on closed subsets to the entire space in standard topological spaces. The theory guarantees the existence of local extensions by skillfully creating them and utilizing the features of normalcy and uniform convergence. Furthermore, continuity is preserved during this process. This proof demonstrates the interplay between topological characteristics and function extension techniques, emphasizing the significance of the theorem in both theoretical and applied mathematics (Fridy & Orhan, 1997).

4.4.2. Implications and Applications of Tietze Extension Theorem

In the field of mathematics, the Tietze Extension Theorem is a fundamental notion that is specifically utilized to extend continuous functions that are defined on closed subsets of normal topological spaces to the entire space. Mathematicians can effortlessly expand the nature of functions beyond their initial boundaries thanks to the theorem, which provides a powerful instrument for analyzing and altering topological spaces (Frantz, 1995).

4.4.3. Use in Various Topological Spaces

Because it may be utilized in a broad variety of topological spaces, the Tietze Extension Theorem is an especially beneficial mathematical concept. When working with compact spaces, metric spaces, or extra abstract topological structures, the theorem ensures that continuous functions that are constructed on closed subsets can be extended while preserving their continuity. This is true regardless of whether the continuous functions are being extended or not. In fields such as analysis, topology, and functional theory, where it is essential to have an understanding of the behavior of functions across several different spatial domains, the versatility of this instrument is of the utmost importance (Spreen, 1998).

Example 1: An examination of functional analysis

The extension of the Tietze Linear functionals that are first given on subspaces of normed vector spaces is typically extended to the entire space by use of theorem, which is a technique that is frequently used in functional analysis. A more comprehensive comprehension of the composition and properties of function spaces is made possible as a result of the utilization of the theorem by mathematicians, who can ensure the unbroken and unchanged nature of linearity in these expansions (Knapp, 2005).

Example 2: The study of topology in algebraic algebra

To construct mappings that preserve basic topological characteristics across different spaces, the theorem in algebraic topology is an extraordinarily important concept. As an illustration, it makes it possible to extend the scope of mappings between homology or cohomology groups that are defined on closed subsets to include the entirety of the space. This is necessary to do an investigation into the inherent properties of topological spaces and the higher-dimensional equivalents of those spaces (Wallman, 1938).

Another illustration of the application of geometric principles in the subject of mathematical physics is provided here.

In the fields of geometry and mathematical physics, the Tietze Extension Theorem is an essential and fundamental concept, particularly when it comes to dealing with the continuous representation of spatial occurrences. The construction of smooth and well-defined functions that appropriately depict sophisticated geometrical structures is made possible as a result of this. By extending functions that are defined on closed subsets, researchers in the fields of mathematics and physics can accurately represent and study geometric configurations and physical systems in several dimensions. In addition to having direct applications in mathematics, the Tietze Extension Theorem has larger repercussions in other fields, such as computer science, where it helps in the construction of algorithms for spatial data processing and geometric modeling. In addition, the theorem has direct applications in geometric modeling. The seamless integration of functions across a wide variety of domains improves the computational efficiency and accuracy of a wide variety of applications, including image processing and machine learning, amongst others (Hyers, 1945).

To provide a concise summary, the Tietze Extension Theorem is a basic notion in modern mathematics that enables the smooth extension of continuous functions across a variety of topological spaces. This notion has applications in a wide variety of domains, ranging from fundamental analysis to complicated theoretical frameworks. This highlights the significance of this concept in terms of its ability to influence the understanding of spatial linkages and functional activity. The theorem continues to be an essential instrument for the investigation and utilization of the intrinsic properties of topological spaces and their applications in both theory and practice, even though the mathematical study is in the process of progress (Frank, 1971).

ACTIVITY 4.1.

Explore the practical applications of the Tietze Extension Theorem in various topological spaces, emphasizing how it extends continuous functions defined on closed subsets to broader domains, enhancing your understanding of function extension in topology.

SUMMARY

This chapter delves into the foundational concepts of separation axioms in topology, which define the degree of separation and closeness between points and sets within topological spaces.

The hierarchy begins with T0 spaces, where distinct points can be distinguished by open sets. Moving up, T1 spaces ensure that every singleton set is closed, allowing for a more refined separation between points. At the pinnacle, T2 spaces, also known as Hausdorff spaces, guarantee the strongest form of separation: every pair of distinct points has disjoint neighborhoods. Hausdorff spaces are particularly important in analysis and geometry where precise distinctions between points are necessary.

Beyond these basic separation axioms, the chapter explores regular spaces, which generalize T1 spaces by ensuring that closed sets and points not in them can be separated by disjoint open sets. Normal spaces extend this concept further by providing a separation of disjoint closed sets using disjoint open neighborhoods. These properties are fundamental in understanding the structure and behavior of topological spaces under various conditions.

A cornerstone theorem discussed in this chapter is Urysohn's Lemma, which plays a crucial role in constructing continuous functions between topological spaces. It states that in a normal space, given any two disjoint closed sets, there exists a continuous function that separates them, illustrating the power of normal spaces in topological analysis.

The chapter concludes with the Tietze Extension Theorem, a powerful tool in functional analysis that extends continuous functions defined on closed subsets to the entire space. This theorem has wide-ranging applications in mathematics and engineering, enabling the seamless extension of functions across topological spaces.

In essence, Chapter 4 elucidates the fundamental principles of separation axioms in topological spaces, highlighting their significance in defining and analyzing the structural properties of spaces and their applications in various branches of mathematics and beyond.

MULTIPLE CHOICE QUESTIONS

1. **Which separation axiom ensures that for any two distinct points in a space, there exist disjoint neighborhoods around each point?**
   a. T0
   b. T1
   c. T2 (Hausdorff)
   d. Regular

2. **Which theorem allows the extension of continuous functions defined on closed subsets of a space to the entire space?**
   a. Urysohn's Lemma
   b. Tietze Extension Theorem
   c. Tychonoff's Theorem
   d. Sierpinski's Theorem
3. **In a T1 space, what can be said about singletons?**
   a. They are open.
   b. They are closed.
   c. They are compact.
   d. They are dense.
4. **What is the key property that distinguishes a normal space from a regular space?**
   a. Every closed set is the intersection of open sets.
   b. Every two disjoint closed sets can be separated by disjoint open sets.
   c. Every point has a neighborhood basis consisting of closed sets.
   d. Every open set is the union of closed sets.
5. **Urysohn's Lemma is primarily used to construct:**
   a. Homeomorphisms between spaces.
   b. Continuous functions from open sets to closed sets.
   c. Continuous functions that separate closed sets.
   d. Borel sets from Lebesgue measure.
6. **Which of the following statements about T2 (Hausdorff) spaces is false?**
   a. Every compact subset is closed.
   b. Limits of sequences are unique.
   c. They are necessarily T1 spaces.
   d. Every singleton set is open.

REVIEW QUESTIONS

1. Define what it means for a topological space to be T2 (Hausdorff). Explain why Hausdorff spaces are particularly useful in topology and analysis.
2. Differentiate between T1 and T2 spaces in terms of the separation properties they guarantee. Provide examples of spaces that are T1 but not T2.
3. What is Urysohn's Lemma? State its formal statement and outline how it is used to construct continuous functions in topological spaces.
4. Explain the significance of regular spaces in topology. How do they extend the separation properties of T1 spaces?

5. Discuss the implications of the Tietze Extension Theorem in topology. Provide a brief proof outline of the theorem and describe its applications.
6. Compare and contrast T0, T1, T2, regular, and normal spaces. Give examples where these concepts are applied in different areas of mathematics or real-world scenarios.
7. These questions aim to assess understanding of the key concepts and theorems covered in the chapter on Separation Axioms.

Answers to Multiple Choice Questions

1. (c) 2. (b) 3. (b) 4. (b) 5. (c)

6. (a)

REFERENCES

1. Al-Shami, T. M. (2021). On Soft Separation Axioms and Their Applications on Decision-Making Problem. Mathematical Problems in Engineering, 2021(1), 8876978.
2. Auscher, P., & Hytönen, T. (2013). Orthonormal bases of regular wavelets in spaces of homogeneous type. Applied and Computational Harmonic Analysis, 34(2), 266–296.
3. Ávila, G., & Antonio, J. (2005). and separation axioms between and. Divulgaciones Matemáticas, 13(2), 90–98.
4. Black, D. R., & Smith, H. A. (1993). Notable exceptions? New and arrested directions in Canadian foreign policy literature. Canadian Journal of Political Science/Revue canadienne de science politique, 26(4), 745–774.
5. Bregman, J. H., Šapirovskij, B., & Šostak, A. (1984). On partition of topological spaces. ⊠asopis pro p⊠stování matematiky, 109(1), 27–53.
6. Butler, I. R., Cullen, W. R., Einstein, F. W., Rettig, S. J., & Willis, A. J. (1983). Synthesis of some ring-substituted [1] ferrocenophanes and the structure of four representative examples. Organometallics, 2(1), 128–135.
7. Carolan, M. (2017). More-than-active food citizens: a longitudinal and comparative study of alternative and conventional eaters. Rural Sociology, 82(2), 197–225.
8. Chaperon, M. (2008). The Lipschitzian core of some invariant manifold theorems. Ergodic Theory and Dynamical Systems, 28(5), 1419–1441.
9. Chedin, F., & Benham, C. J. (2020). Emerging roles for R-loop structures in the management of topological stress. Journal of Biological Chemistry, 295(14), 4684–4695.
10. Ciesielska, M., & Ciesielski, K. C. (2017). Differentiable extension theorem: a lost proof of V. Jarník. Journal of Mathematical Analysis and Applications, 454(2), 883–890.
11. Cigler, B. A. (1999). Pre-conditions for the emergence of multicommunity collaborative organizations. 1. Review of Policy Research, 16(1), 86–102.

12. Clarke, E., Grumberg, O., Jha, S., Lu, Y., & Veith, H. (2000). Counterexample-guided abstraction refinement. In Computer Aided Verification: 12th International Conference, CAV 2000, Chicago, IL, USA, July 15–19, 2000. Proceedings. Vol. 12, (pp. 154–169).

13. Coley, J. D., & Tanner, K. D. (2012). Common origins of diverse misconceptions: Cognitive principles and the development of biology thinking. CBE—Life Sciences Education, 11(3), 209–215.

14. Connor-Greene, P. A. (1993). The therapeutic context: Preconditions for change in psychotherapy. Psychotherapy: Theory, Research, Practice, Training, 30(3), 375.

15. DeLozier, S. J., & Rhodes, M. G. (2017). Flipped classrooms: A review of key ideas and recommendations for practice. Educational psychology review, 29(1), 141–151.

16. Di Donato, D., Fässler, K., & Orponen, T. (2022). Metric Rectifiability of H-regular Surfaces with Hölder Continuous Horizontal Normal. International Mathematics Research Notices, 2022(22), 17909–17975.

17. Dritschel, M. A., & Rovnyak, J. (1996). Operators on indefinite inner product spaces. Lectures on operator theory and its applications, 3, 141–232.

18. Egenhofer, M. J., & Herring, J. (1990). Categorizing binary topological relations between regions, lines, and points in geographic databases. The, 9(94–1), 76.

19. Estrada, E., & Uriarte, E. (2001). Recent advances on the role of topological indices in drug discovery research. Current Medicinal Chemistry, 8(13), 1573–1588.

20. Faure, C. A. (2002). An elementary proof of the fundamental theorem of projective geometry (dedicated to Alfred Frölicher). Geometriae Dedicata, 90(1), 145–151.

21. Fekete, I., Gregorics, T., Pusztai, K. K., & Veszprémi, A. (2019). Programming theorems and their applications. Teaching Mathematics and Computer Science, 17(2), 213–241.

22. Frank, M. J. (1971). Probabilistic topological spaces. Journal of Mathematical Analysis and Applications, 34(1), 67–81.

23. Frantz, M. (1995). Controlling Tietze-Urysohn extensions. Pacific Journal of Mathematics, 169(1), 53–73.

24. Fridy, J. A., & Orhan, C. (1997). Statistical core theorems. Journal of Mathematical Analysis and Applications, 208(2), 520–527.

25. Goldfarb, I., Gol'dshtein, V., & Maas, U. (2004). Comparative analysis of two asymptotic approaches based on integral manifolds. IMA journal of Applied Mathematics, 69(4), 353–374.

26. Gomez-Zwiep, S. (2008). Elementary teachers' understanding of students' science misconceptions: Implications for practice and teacher education. Journal of Science Teacher Education, 19(5), 437–454.

27. Grabiner, S. (1986). The Tietze extension theorem and the open mapping theorem. The American Mathematical Monthly, 93(3), 190–191.

28. Heldermann, N. C. (1980). The category of ⊠-completely regular spaces is simple. Transactions of the American Mathematical Society, 262(2), 437–446.

29. Holmes, R., Walker, R. J., & Murphy, G. C. (2006). Approximate structural context matching: An approach to recommend relevant examples. IEEE Transactions on Software Engineering, 32(12), 952–970.

30. Huttenlocher, D. P., Klanderman, G. A., & Rucklidge, W. J. (1993). Comparing images using the Hausdorff distance. IEEE Transactions on pattern analysis and machine intelligence, 15(9), 850–863.

31. Hyers, D. H. (1945). Linear topological spaces. Bulletin of the American Mathematical Society, 51(1), 1–21.

32. Jenavee, K. S., Asokan, R., & Nethaji, O. (2022). NEW SEPARATION AXIOMS IN. Turkish Journal of Computer and Mathematics Education (TURCOMAT), 13(03), 133–139.

33. Kallio, K. P. (2018). Not in the same world: topological youths, topographical policies. Geographical Review, 108(4), 566–591.

34. Kaplansky, I. (1947). Lattices of continuous functions. Bulletin of the American Mathematical Society, 53(6), 617–623.

35. Karaisa, A., & Başar, F. (2014, August). Some new paranormed sequence spaces and core theorems. In AIP Conference Proceedings. American Institute of Physics. Vol. 161(1), (pp. 380–391).

36. Kernighan, B. W., & Plauger, P. J. (1974). Programming style: Examples and counterexamples. ACM Computing Surveys (CSUR), 6(4), 303–319.

37. Knapp, A. W. (2005). Topological Spaces. Basic Real Analysis: Along with a companion volume Advanced Real Analysis, 441–484.

38. Kopperman, R. D., & Tkachuk, V. V. (2003). The approximation of compacta by finite T0-spaces. Quaestiones Mathematicae, 26(3), 355–370.

39. Kreinovich, V. (2010). Metrization Theorem for Space-Times: From Urysohn's Problem Towards Physically Useful Constructive Mathematics. Fields of Logic and Computation: Essays Dedicated to Yuri Gurevich on the Occasion of His 70th Birthday, 470–487.

40. Kubiak, T. (1987). L-fuzzy normal spaces and Tietze extension theorem. Journal of mathematical analysis and applications, 125(1), 141–153.

41. Kutbi, M. A., Alam, A., & Imdad, M. (2015). Sharpening some core theorems of Nieto and Rodríguez-López with application to boundary value problem. Fixed Point Theory and Applications, 2015, 1–15.

42. Lancaster, P. (1977). A fundamental theorem on lambda-matrices with applications—I. Ordinary differential equations with constant coefficients. Linear Algebra and Its Applications, 18(3), 189–211.

43. Landry III, R. J. (2024). Applicability of Discharge Exceptions to Corporate Debtors in Subchapter V: A "Death Blow" to Rescuing Small Businesses. The Journal of Business, Entrepreneurship & the Law, 17(1), 1–36.

44. Lassez, J. L., & Marriott, K. (1987). Explicit representation of terms defined by counter examples. Journal of Automated reasoning, 3, 301–317.

45. Lin, Q., Mao, R., Liu, J., Xu, F., & Cambria, E. (2023). Fusing topology contexts

and logical rules in language models for knowledge graph completion. Information Fusion, 90, 253–264.

46. Ling, G. C. L., Shahrill, M., & Tan, A. (2016). Common misconceptions of algebraic problems: Identifying trends and proposing possible remedial measures. Advanced Science Letters, 22(5–6), 1547–1550.
47. Luchko, Y. (2020). Fractional derivatives and the fundamental theorem of fractional calculus. Fractional Calculus and Applied Analysis, 23(4), 939–966.
48. Ludwig, A. W., & Affleck, I. (1994). Exact conformal-field-theory results on the multi-channel Kondo effect: Asymptotic three-dimensional space-and time-dependent multi-point and many-particle Green's functions. Nuclear Physics B, 428(3), 545–611.
49. Miñana, J. J., Šostak, A., & Valero, O. (2023). On metrization of fuzzy metrics and application to fixed point theory. Fuzzy Sets and Systems, 468, 108625.
50. Morsi, N. N. (1991). The Urysohn Lemma for fuzzy neighbourhood spaces. Fuzzy Sets and Systems, 39(3), 347–360.
51. Mursaleen. (2004). Almost strongly regular matrices and a core theorem for double sequences. Journal of Mathematical Analysis and Applications, 293(2), 523–531.
52. Neiskashova, E. V., & Fomina, T. N. (2020, November). Counterexamples and critical thinking formation. In Journal of physics: conference series. Vol. 1691(1)1, pp. 012232).
53. Nyumba, T., Wilson, K., Derrick, C. J., & Mukherjee, N. (2018). The use of focus group discussion methodology: Insights from two decades of application in conservation. Methods in Ecology and evolution, 9(1), 20–32.
54. Pąk, K. (2014). Tietze extension theorem for n-dimensional spaces. Formalized Mathematics, 22(1), 11–19.
55. Pedersen, G. (1997). A strict version of the non-commutative Urysohn lemma. Proceedings of the American Mathematical Society, 125(9), 2657–2660.
56. Radulescu, M., & Radulescu, S. (1980). Global inversion theorems and applications to differential equations. Nonlinear Analysis: Theory, Methods & Applications, 4(5), 951–965.
57. Rodriguez, M., Delon, J., & Morel, J. M. (2018). Covering the space of tilts. application to affine invariant image comparison. SIAM Journal on Imaging Sciences, 11(2), 1230–1267.
58. Shavitt, Y., & Weinsberg, U. (2011). Quantifying the importance of vantage point distribution in internet topology mapping (extended version). IEEE Journal on Selected Areas in Communications, 29(9), 1837–1847.
59. Siddiqui, S., Shikotra, A., Richardson, M., Doran, E., Choy, D., Bell, A.,... & Bradding, P. (2018). Airway pathological heterogeneity in asthma: visualization of disease microclusters using topological data analysis. Journal of Allergy and Clinical Immunology, 142(5), 1457–1468.
60. Skånér, Y., Bring, J., & Strender, L. E. (2005). Selecting representative case vignettes for clinical judgement studies: Examples from two heart failure studies. Journal of Symbolic Logic Quality and quantity, 38, 621–635.

61. Soltanian-Zadeh, H., & Peck, D. J. (2001). Feature space analysis: Effects of MRI protocols. Medical physics, 28(11), 2344–2351.

62. Spreen, D. (1998). On effective topological spaces. The Journal of Symbolic Logic, 63(1), 185–221.

63. Sylvester, J. J. (1878). XXV. Proof of the hitherto undemonstrated fundamental theorem of invariants. The London, Edinburgh, and Dublin Philosophical Magazine and Journal of Science, 5(30), 178–188.

64. Tachtsis, E. (2019). The Urysohn lemma is independent of ZF+ Countable Choice. Proceedings of the American Mathematical Society, 147(9), 4029–4038.

65. Wallman, H. (1938). Lattices and topological spaces. Annals of Mathematics, 39(1), 112–126.

66. Yamashita, M. (1992). Melnikov vector in higher dimensions. Nonlinear Analysis: Theory, Methods & Applications, 18(7), 657–670.

CHAPTER

5

Countability

LEARNING OBJECTIVES

After studying this chapter, you will be able to:

1. Understand the fundamental concepts and principles of countability in topology.
2. Explore and apply foundational theorems related to countable sets.
3. Analyze elementary examples and their significance in the context of countability.
4. Investigate advanced examples and understand their impact on the broader study of topology.
5. Examine the concept of universal countability and its theoretical implications.
6. Develop the ability to apply rigorous definitions and formalisms to problems involving countability.

KEY TERMS FROM THIS CHAPTER

- Countable Set
- Countable Basis
- First Countability Axiom
- Lindelöf Space
- Uncountable Set
- Countable Subcover
- Second Countability Axiom

UNIT INTRODUCTION

In topology, like in other fields of mathematics, one encounter numerous infinite sets. Yet, one will come to realize, certain infinite sets have different sizes. Countably infinite sets, although infinite, can be considered relatively small in a specific way. Interestingly, these sets can be considered as the "smallest infinite sets." Working with countable sets is advantageous due to the ability to list their elements, which facilitates tasks such as inductive proofs (Deal, 2017).

In the preceding part, one acquired knowledge that the set Q, which consists of rational numbers, exhibits density within the set R. The next part will provide proof that shows Q, the set of rational numbers, is countable. It is advantageous because while R itself is a vast assemblage (innumerable), there is a finite subset inside it that is "close to everything," according to the conventional topology (De Belder, 2011). Similarly, the standard topology on R consists of a multitude of sets (a large array of sets), yet as one already demonstrated in a previous question without explicitly stating it, this topology possesses a basis that can be counted, indicating that nearly all the details about the topology are defined by a "limited" assortment of open sets. This statement does not hold on to the Sorgenfrey Line, as one will discover later (Wicke & Worrell, 1976).

In this course, the idea of countability is frequently discussed in the context of topology. These notes aim to ensure that students fully understand the related concepts before using terms like "countable" along with "uncountable" extensively (Lonti & Gregory, 2007).

5.1. OVERVIEW OF COUNTING

The topic of countability along with uncountability revolves around the "sizes" of sets and the comparison of these sizes. This is something that may be easily overlooked when working with finite sets (Cappo et al., 2006).

For instance, consider a scenario where there is a room using seven individuals and a set of seven hats. How can one verify if the number of hats matches the number of people? It is most probable that one will count the hats first, followed by tallying the individuals. In both instances, the answer would be 7. It is evident that 7 equals 7, leading to the conclusion that the quantity of hats matches the quantity of people. Indeed, one must be right (Gauffriau, 2021).

Now picture a scenario where someone requested one to perform this task, but one was unaware of the concept of quantifying objects or individuals with numbers. Is it possible to demonstrate that the quantity of hats matches the quantity of people?

Remember

Remember the foundational concepts of countability in mathematics, which involve understanding the different types of sets and their cardinalities, particularly focusing on countable sets and their properties.

One idea is to distribute hats to people by placing one on every person's head via collection. After finishing the task, it is crucial to confirm two aspects. First, ensure that all hats are utilized. Secondly, it is important to make sure that all has a hat. When both claims happen to be true, one could reasonably conclude that the quantity of hats is equivalent to the quantity of people. It can be argued that if there is some hats left after everyone has one upon their head, then there will be an excess of hats in comparison to the number of people. However, if there remain people who decide not to put on hats, it implies that there is a shortage of hats compared to the amount of people (Bakker et al., 2015).

At first glance, it may seem unimportant, but it's important to recognize that the technique described allows for determining the equality of two sets lacking counting. Even without any understanding of numbers, for this reason. This concept involves evaluating the sizes of collections, which can be expanded towards infinity (Van Den et al., 2003).

When dealing with infinite sets, the process of comparing their dimensions is not as straightforward as counting utilizing integers. In that situation, the strategy provided in the example above is the only option for comparing sets.

In a more numerical context, it is interesting to consider that actions involving people and hats can be viewed as a function which maps a group individuals to the set of hats. This function could be described through

g(Person X) = the hat one put on Person X's head.

If each person is given only one hat, then g seems like an injective. When a hat is placed on every person, it becomes a surjective. If g occurs to be both injective as well surjective (i.e., bijective), then it can be concluded that the sets for hats as well as individuals have the same number of elements (Liu et al., 2019).

This concept is utilized to evaluate the magnitudes of various collections.

Definition 2.1. When considering a pair of set, A and B, it is possible to determine if they have identical a cardinality through finding a bijection $f : A \to B$. This is usually shown as $|A| = |B|$ (Patwal et al., 2023).

For those who enjoy the elegance of precise mathematical language, it can be said that two sets of equal size meet (Bai et al., 2022).

Definition 2.2. When comparing a pair of sets for numbers A or B, one can figure out if A has a smaller or equal number of elements than B by checking if there is a one-to-one mapping $f : A \to B$ or a mapping (surjection) $g : B \to A$. The usual way to express this is via $|A| \leq |B|$.

If there exists an injective function $f : A \to B$ yet there is no surjective function between A to B, one defines A to have less cardinality than B, denoted as $|A| < |B|$ (Saghir & Lin, 2015).

Theorem 2.3 (Cantor-Schr¨oder-Bernstein Theorem). Consider A, B *be sets. When* $|A| \leq |B|$ *or*

$|B| \leq |A|$, *thus* $|A| = |B|$.

This theorem is far more complex than it appears. If there exists a injection $f : A \to B$ with the injection $g : B \to A$ that means there exists a bijection $h : A \to B$ Attempt to show it on their own, as onee will discover that it is hard, even for sets that have a little number of elements. One suggests examining the proof for that theorem, since it is both elegant and serves as a fundamental illustration of a valuable proof strategy known as a "back and forth argument" (Ribeiro et al., 2021).

5.2. FOUNDATIONS OF COUNTABILITY

Definition 3.1. If $|A|=|N|$, Set A has a countably infinite number of elements; otherwise, it is just countable.

$|A|\leq|N|$.

A set becomes countable if it is equivalent to that of a subset for natural numbers. One uses the term "countable" to refer to a set that is actually "countably infinite," assuming it a set in question is infinite. It is important to understand that the empty set $\varnothing$ can be counted, as there exists an equation $f:\varnothing\to N$ which is an injection by vacuity (Sanders, 2022).

The dominant notion here is that a collection is countable if its constituents can be enumerated. Ultimately, an injection $f:A\to N$ is just a method of giving a number for every member of A in a logical manner, ensuring every element of A is assigned a unique number. A set A is considered countably infinite if it is possible to create a list of its elements (Kechris, 1975).

$a_1, a_2, a_3, a_4, a_5, a_6, \ldots$

Prove that each element of (a) seems on the list only once and that the indicates each component of A. (With this list, the equation given by $f(n)=a_n$ seems a one-to-one correspondence between $N\to A$.) Our proofs for the countability of sets will often be presented as lists (Makowsky, 1978).

It is important to emphasize that this way of evaluating sizes can produce results which may appear unconventional to individuals that are used to thinking in terms of sizes for finite sets. Take, for instance, a clear fact regarding finite sets: When A is a finite set and B represents a proper sub set of A, it is evident that B is lesser than A. Essentially, when one start from an infinite set A or remove certain elements, what remains is less than A. This does not apply to infinite set (Morley, 1970).

Example 3.2. Suppose $E\subseteq N$ is set of even numbers. Thus $|E|=|N|$. Therefore:

$2, 4, 6, 8, 10, 12, 14, 16, \ldots$

5.3. ELEMENTARY EXAMPLES AND FUNDAMENTAL THEOREMS

Let explore some simple instances of sets that can be counted. By examining these examples, one can uncover some basic truths about sets that can be counted (Baer, 1944).

Example 4.1. This set $A = \{n \in N : n > 7\}$ is countable. One can enumerate the elements using a bijective manner:

$8, 9, 10, 11, 12, 13, \ldots$

or suppose of the bijection $f : N \rightarrow A$ represented by $f(n) = n + 7$.

It is evident that there is nothing exceptional regarding the number 7 in this context. Initially, one began with a set which could be counted infinitely, then eliminated a finite number of items, resulting in a set that remained countably infinite. By removing the initial seven elements, one was able to clarify f in a handy manner, but this step was not obligatory (Eisermann, 2012).

Proposition 4.2. If set A seems countable whereas set B, which is a subset of A, is finite, so the set difference $A \setminus B$ is also countable.

Evidence. The conclusion is self-evident when A is finite, hence one will focus on the case where A becomes countably infinite (Zanini et al., 2017).

Explaining the concept of this proof is simpler than putting it into writing. To produce a list for the elements in $A \setminus B$, begin using a list of an elements in A. Remove the elements in B via the list, and then compress all to fill within the gaps left by the changes (Post, 1921).

As a demonstration, he specifically addresses the scenario where B offers two elements. Assume that A is countable and let $f : N \rightarrow A$ be a bijection. Now, consider the set $B = \{f(17), f(2523)\}$. Then a function $g : N \rightarrow A \setminus B$ is a bijection.

$$g(n)=\begin{cases} f(n) & \text{if} n<17 \\ f(n+1) & \text{if} 17\le n<2522 \\ f(n+2) & \text{if} n\ge 2522 \end{cases}$$

Did you know?

Did you know that one of the fundamental theorems related to countability is Cantor's theorem, which states that the set of real numbers is uncountable, contrasting with countable sets like the set of integers?

Verify the bijection within the sets N along with $A \setminus B$, while familiarizing oneself with the process of defining a function in a broader sense.

One may use the same concept, but in reverse. That is, assuming one begins with a set that can be counted and then introduces a finite number of additional elements, the outcome will still be a set that can be counted (Huntington, 1906).

Example 4.3. The set $A=\{n\in Z : n\geq -7\}=N\cup\{-7,-6,\ldots,-1,0\}$ is countable. It is evident within the function $f:N\to A$ defined by $f(n)=n-8$.

Proposition 4.4. *Let A be counted, and B a set that has a limited number of elements. Following that A ∪ B can count.*

Proof. These results becomes evident if A was finite, so now assume that it is infinitely countable. In addition, one can assume both A and B do not overlap, as the overlap could reduce the size of B and be advantageous for our purpose. Formally,

$A\cup B=A\cup(B\setminus A)$, along with $B\setminus A$ might be finite Thus, it is sufficient to consider sets that lack disjoint sets A and B (Charosh, 1944).

Suppose $f:N\to A$ Confirm that A as countable with enumerate $B=\{b1,b2,\ldots,bk\}$ for k.

One might do this as B is finite. Let define:

$$g(n)=\begin{cases} b_n & \text{if} n\le k \\ f(n-k) & \text{if} n>k \end{cases}$$

When it comes to lists, one essentially adding the components of B for the front of original elements list in A, resulting in $A\cup B$ being shown as

$$b_1,b_2,\ldots,b_k,f(1),f(2),f(3),f(4),\ldots,$$

Next, the updated set will be revised with the letter *"g."*

Therefore, the act of including or excluding a finite number of elements has no impact on the concept of countability. Yet, in the section prior (Example 4.1), it was shown that in certain cases, it is possible to eliminate an infinite number of elements from a measurable set and still retain a countable set (Greenberg, 2006). Yet again:

Example 4.5. Consider $E\subseteq N$ is the set of positive even numbers. It can be observed that E is countable, in evidenced by

$$f(n)=2n.$$

Yet again, this instance allows one to derive a set of broader data.

Proposition 4.6. Consider A as a set that has an infinite number of elements, while B is a subset of A that is also infinite. Then B can be counted.

Proof. In this instance, the concept is like Proposition 4.2. It is necessary to make a list of the components of A, eliminate any elements that do not fall within the scope of B, and then condense everything to occupy all available space. One will ensure this proves more accurate (Artin & Mumford, 1972).

So let $f : N \to A$ be a bijection that demonstrates that set A is countable. One aims to create a one-to-one correspondence $g : N \to B$.

The subsequent proof essentially entails the following: Enumerate the items of set A, then select the element from set B with the lowest index in this enumeration and designate it as g (1). Next, select the part for B with the subsequent smallest index and designate it as g (2). Iterate this procedure inductively, assigning g(k) as the member of B that has the kth smallest index. Next, one shows that g serves as a bijection (Schmidtke & Kuperman, 2017).

Now $k_1 = min k \in N : f(k) \in B = min(f^{-1}(B))$. Which is, k_1 like smallest number which gets mapped through B by f. Define $g(1) \square f(k_1)$. One proceed inductively.

Suppose one defines $g(1), g(2), \ldots, g(n)$. consider

$$k_{n+1} = min k \in N : f(k) \in B\, g(1), \ldots, g(n).$$

In this case, k_{n+1} shows the lowest value that f assigns to every element in B which is still not seen by g. (B forms an infinite set, $B \setminus \{g(1), \ldots, g(n)\}$ remains infinite and, importantly, nonempty. Therefore, the existence of this minimum value is guaranteed (Furuta, 2008).

Therefore, let $g(n+1)$ be defined as $f(k_{n+1})$. By proceeding in this manner, one establishes an equation g: $N \to B$ that is injective. To demonstrate subjectivity, choose an element b within the set B and assume, for the sake of contradiction, that b is not associated with the range with the function g.

$$X = \{n \in N : f(n) \in B\, but\, f(n)\, is\, not\, in\, the\, range\, of\, g\}.$$

That is, X represents the set for indices where g does not include elements from B. After that, X set is not empty, as $f^{-1}(b)$ belongs to X at least. As a not empty subset out of naturals, X requires a minimum element, that one refer to as n_0. However, it follows from the definition for ki that n_0 is a k_N at least $N \leq n$.

To better understand this, it is important to observe that a maximum of n0 of the k_i's may happen before n_0. Let M represent the number of incidences (Mustafa, 2020). In short,

M = |{1, 2, 3,... n_{-0}} ∩ { k_i : i ∈ N }|

Which means $k_1, k_2, \ldots, k_M$ are each $\leq n_0$ having no other k_i's below n_0. But:

$$n_0 = minn \in N : f(n) \in Bg(1), \ldots, g(M).$$

Thus, by defining k_i's, one has $n_0 = k_{M+1}$. So M + 1 is the N one explained earlier.

This is a paradox, as $g(n_0) = g(k_N) \in B$ according to the definition of g, yet n_0 an element of X. Based on analysis, it can be inferred that X is empty, which implies g is surjective.

Another viewpoint on these results is to view it as a form of dichotomy theorem. When a subset is taken from a countably infinite set, it can only be both finite or infinite. This statement asserts that when the subset cannot be finite, it needs to be countably infinite. Put simply, there are no intermediate sizes that allow a set to be positioned between being finite along with countably infinite (Lachlan,1984).

Like the two "corresponding" propositions one previously demonstrated, Proposition 4.6 also yields another result when reversed. Put simply, if one begins with a set that may be counted and then add a finite or infinite number of elements to it, the resulting set remains countable (Delbaen & Schachermayer, 1994).

Proposition 4.7. *Suppose A, B represents countable sets. Thus A ∪ B is countable.*

Proof. If A and B were finite, this is clear. If only one of the two is finite, this corresponds to Proposition 4.4. Let assume that both are infinitely countable. Once more, it is safe to assume that A and B do not overlap, without sacrificing the overall applicability of the argument.

One may do this because $A \cup B = A \cup (B \setminus A)$, and according to the prior proposition, $B \setminus A$ must be both finite as well as countably infinite (Heathcote, 2021).

Suppose $f : N \to A$ or $g : N \to B$ provide bijections that demonstrate that sets *A and B* are now countable.Defination

$h : N \to A \cup B$ as:

$$h(n) = \begin{cases} f(k) & \text{if} n = 2k \\ g(k) & \text{if} n = 2k - 1 \end{cases}$$

In the realm of lists, combine the elements of both *A and B* by integrating the two lists (Ball, 1997).

Be confident in the fact that h has a bijection

Corollary 4.8. Z *is countable.*

Proof. Note that $Z = (N \cup \{0\}) \cup (-N)$, whereas −N seems abuse of notation representing $\{-n : n \in N\}$, along with negative integers. The result can be deduced from Propositions 4.7 or 4.4.

Exercise 4.9. Create a clear one-to-one correspondence $f : N \to Z$.

Exercise 4.10. Show that an infinite union of sets, each of which is countable, is also countable. (Hint: Utilize the prior Proposition along with the method of induction).

Now, let's introduce a proposition which will give us a little more flexibility. All the proofs we have done so far to show an object is countably infinite rely on bijections to N. Bijections are quite elegant functions, since they possess inverses that are also bijections. Additionally, the composition between two bijections results in another bijection (Gulgowski et al., 2021).

Proposition 4.11. Consider A as an infinite set, while C remains a fixed countably infinity set. Then the following statements are all the same. *A is countable.*

1. *There exists a bijection* $f : N \to A$.
2. *There exists a bijection* $g : A \to N$.
3. *There exists a bijection* $h : A \to C$, *or* $h : C \to A$.
4. *There exists an injection* $i : A \to N$, *or* $i : A \to C$.
5. *There exists a surjection* $s : N \to A$, *or* $s : C \to A$.

Proof. **Exercise.** (Almost anything here can be derived from fundamental attributes of functions, or one has earlier proven.) It may be necessary to put in some effort to demonstrate that (6) is equivalent of the others.

Exercise 4.12. In a subsequent part of this course, one will acquire knowledge on the Axiom for Choice. Once one had acquired knowledge about it (or at present if already acquainted with it), endeavor to identify whether any of the connections between these facts are contingent upon a certain degree of decision-making (Mason, 2001) (Figure 5.1).

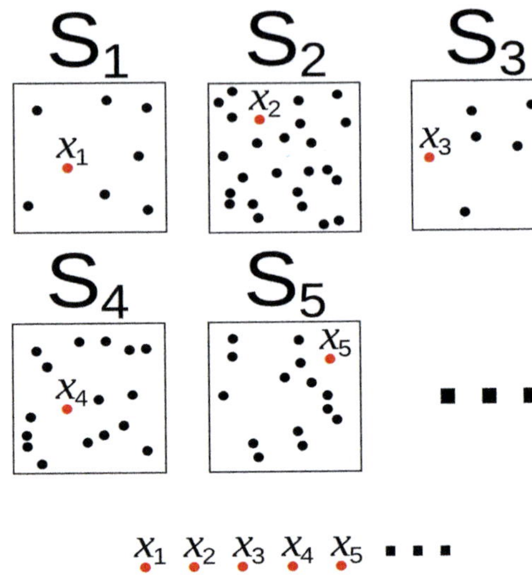

Figure 5.1. Each set in the countable sequence of sets (Si) = S1, S2, S3,... contains a non-zero, and possibly infinite (or even uncountably infinite), number of elements. The axiom of countable choice allows us to arbitrarily select a single element from each set, forming a corresponding sequence of elements (xi) = x1, x2, x3... (Source: Dcoetzee, Creative Commons License)

5.4. ADVANCED EXAMPLES AND SIGNIFICANT THEOREMS

Previously, one demonstrated the peculiar phenomenon that infinite sets may have the same cardinality as their proper subsets. One discovered as N and Z had equal magnitudes, despite the common perception that Z was at least twice as large as N. Indeed, there exist two flawless replicas of N within Z. This is peculiar.

In this section, one shall demonstrate that N is equipotent to a set that contains an endless number of exact replicas of itself (Wang, 1960).

Theorem 5.1. N × N seems *countable.*

Proof. There are a couple of ways to demonstrate this. One approach involves creating a visual representation and indicating specific elements. The other method entails clearly defining a one-to-one correspondence function $f:N\times N\to N$. The previous method is the most optimal approach (Weber, 2004) (Figure 5.2).

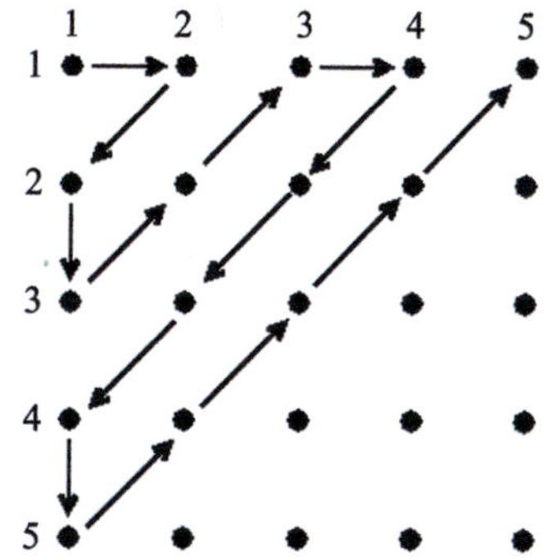

Figure 5.2. Be convinced that there is a proof which N × N was countable (Source: College Math Teaching, Creative Commons License)

To provide a clearer proof, now describe a function $f:N\times N\to N$ by

$f(n,m)=2^n3^m$.

This function remains injective due to the distinct nature for prime decompositions, and therefore N×N is countable according to Proposition 4.11, part 5 (Figure 5.3).

Corollary 5.2. *Consider A and B sets that can be counted. Then A × B can be classified as countable.*

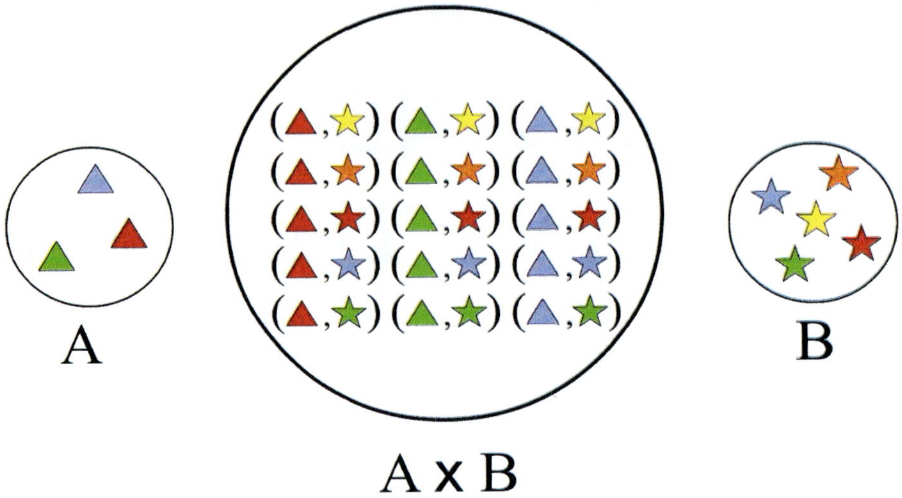

Figure 5.3. If Set A and Set B are countable then A X B is countable (Source: BYJU's, Creative Commons License)

Exercise 5.3. Demonstrate that the Cartesian result of a finite number of sets that can be counted is also a set that can be counted. This can be shown in two distinct manners. It is important to understand the limitations of both proofs when applied to goods from countable sets (Neubrand, 1989).

Corollary 5.2 leads to a crucial outcome in this section that is highly relevant to our objectives.

Corollary 5.4. Q *is countable.*

Proof. It is evident that the set Q stands infinite. To demonstrate that it is countable, one must establish an injection from Q toward set Z × N, that is known to be countable according to the prior conclusion (Fekete et al., 2019).

It should be noted that Q may be represented as

$$\frac{p}{q}: p \in \mathrm{Z},\ q \in \mathrm{N}, \text{ and fraction shows lowest terms}$$

Defining a function $f: \mathrm{Q} \to \mathrm{Z} \times \mathrm{N}$ by $f\left(\frac{p}{q}\right) = (p, q)$. This map is an injective function (verify this!) and, as a result, Q is countable according to Proposition 4.11, part 5.

Throughout the program, one will extensively discuss this truth in several circumstances. Currently, please pause briefly to fully accept and understand the situation. So one has probably been imagining countable sets as sparsely populated collections of points, such as N and Z. On the contrary, Q is highly compact. Everywhere one observes within the set of real numbers, an unlimited number of rational numbers can be found. However, the collection of all rational numbers can be enumerated (Rav, 1999).

The following statement is another consequence of Theorem 5.1. By now one must be overly astonished by it, but it might have been unexpected when we first started discussing countability.

Corollary 5.5. *A union of sets that are countable is also countable. Which is, when* $A_n, n \in N$ *are countable sets, thus* $U_{n \in N} A_n$ *is countable.*

Proof. As previously stated, it can be assumed that a set is each infinite as well mutually disjoint.

Suppose $fn : N \rightarrow An$ bieng a bijection witnessing where A_n is countable, thus defining g : N × N →

$U_{\in N} A_n$ through $g\ (n,\ i) = f_n(i)$ (Figure 5.4).

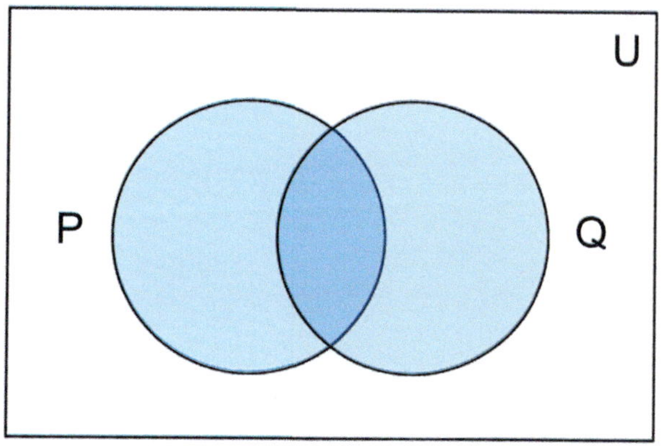

Figure 5.4. If set P and Q are countable then their union is also countable (Source: BYJU's Creative Commons License)

Exercise 5.6. Demonstrate the function g stated above constitutes a bijection, or therefore establish that the set $U_{n \in N} A_n$ remains countable (Kreisel, 1958).

Exercise 5.7. Demonstrate property of set $B_Q = (a, b) \subseteq R : a, b \in Q$ is countable. Which means

R_{usual} is a basis that can be counted (as demonstrated in Big List 2.2, B_Q seems a basis for R_{usual}).

This characteristic of having a basis that can be counted is quite valuable (and remains consistent, as one will explore further) to warrant a designation. One will complete that task at the conclusion of this note (Ma, 2008).

In conclusion, let recap what one has discovered up until now:

i. A subset of a set that can be counted is neither finite nor can be counted infinitely.

ii. Unions for countable sets, whether they are finite or countably infinite, are countable.

iii. Finite Cartesian product sets that may be counted are further able to be counted (Wagner, 1977).

5.5. UNIVERSAL COUNTABILITY: AN INVESTIGATION

Definition 6.1. *A countless set A is one that cannot be counted.*

At this stage, it should come as no surprise (Lecomte, 2020).

Proposition 6.2. *When A is uncountable, B is a set, or* $f: A \rightarrow B$ *is a bijection, thus B seems uncountable.*

Proof.

Alright, which sets are in fact uncountable? There are quite a few of them. One will provide an illustrative example and discuss its implications, along with a technique for generating an unlimited number of sets that cannot be counted.

This proof is widely known and highly regarded, originally presented by Cantor around the 1880s. It was unexpectedly surprising when it happened (Deal, 2016). It is commonly referred to as "Cantor's Diagonalization Proof."

Theorem 6.3 (Cantor). $(0,1) \subseteq R$ *is uncountable.*

Proof. This can be demonstrated through a proof from contradiction. Without a doubt, it is clear that (0, 1) is not finite. Therefore, let's assume that (0, 1) was countably infinite and establish the bijection $f: N \rightarrow (0,1)$ supports this assumption. All real numbers between 0 and 1 have an infinite decimal expansion into a specific form (Chater & Vitányi, 2003).

$$a.a_1a_2a_3a_4a_5a_6a_7\ldots$$

(where we consider a real number that has the decimal expansion which ends with zeros). By following this approach, one can ensure that every single digit found within a decimal expansion of any real number upon our list (the list provided by *f*) will be indexed. As an $x \in (0, 1)$, it is assumed that there exists a single $n \in$ N where $f(n) = x$. Alright, now get down to writing (Tsang, 2017).

$$x = (x_n)_{n=1}^{\infty} = (x_1, x_2, x_3, x_4, x_5, x_6, x_7, \ldots)$$

Picture within mind a table that stretches endlessly, as if it has no boundaries:

$$
\begin{array}{lllllll}
f(1) = 0. & x_1^1 & x_2^1 & x^1 & x_4^1 & x_5^1 & \cdots \\
f(2) = 0. & x_1^2 & x_2^2 & x^2 & x_4^2 & x_5^2 & \cdots \\
f(3) = 0. & x_1^3 & x_2^3 & x_3^3 & x_4^3 & x_5^3 & \cdots \\
f(4) = 0. & x_1^4 & x_2^4 & x_3^4 & x_4^4 & x^4 & \cdots \\
f(5) = 0. & x_1^5 & x_2^5 & x_3^5 & x_4^5 & x_5^5 & \cdots \\
\vdots & \vdots & \vdots & \vdots & \vdots & \vdots & \ddots
\end{array}
$$

one has highlighted the diagonal components within the intriguing portion of the table within red to draw focus on them (Alkazwini, 2016).

Now, create a real number $y{=}0.y\_1y\_2y\_3y\_4y\_5$, which cannot be included in this list (i.e., it cannot belong to the range of *f*). This contradicts belief that *f* acts as a bijection (Lima & Rothstein, 2020).

Now supposes, $y_1 = x_1^1 - 1 -$ 1, instead $x_1^1 = 0$ whereas consider $y_1 = 9$. Which is, let y_{-k} = x_k^k − 1, unless x_k^k = 0, where let $y_k = 9$.

Additionally, by the conclusion of this process, it is imperative to ensure that one has not constructed y = 0 = 0.0000 or y = 1 = 0.9999 So in the end, one simply need to make a modification to one digit which guarantees this (Schmidtke & Kuperman, 2017).

I contend that the values $y = 0, y_1y_2y_3y_4y_5$ do not fall within the range of function *f*.

To illustrate this, assume it was within the image of *f*. Then there must be a natural number n such that $f\,(n)$= y. However, due to the way y is constructed, the nth decimal places of y and f(n) differ at n^{th} $n \in$ N decimal place of $f\,(n)$ (Darji et al., 2019).

This statement presents a contradiction, indicating that y does not appear in the image of *f*. Consequently, it can be concluded that *f* is not a bijection. Given the arbitrary choice of the initial value, it is evident that a bijection between N and (0, 1) cannot exist (Kandi⊠ & Taylor, 2018).

Impressive! I hope that even the most experienced math experts reading will appreciate the elegance of this outcome. Both N and (0, 1) are now infinite, yet (0, 1) has a greater cardinality! Clearly, R cannot be counted (Bogopolski & Zastrow, 2012).

Corollary 6.4. R *is uncountable.*

Proof. All one has to do is demonstrate a one-to-one correspondence h: (0, 1) → R.

$f\,(x)$ = π(x –, $\frac{1}{2}$) represents a bijection among (0, 1) or $(-\frac{\pi}{2}, \frac{\pi}{2})$ (check this!), and $g(x)$ = tan(x), which restricted to $(-\frac{\pi}{2}, \frac{\pi}{2})$, is a one-to-one correspondence between this set and R. The layout $h := g \circ f : (0,1) \to R$ is thus, a bijection, as needed (Gabbay & Mathijssen, 2009).

One finds sets that cannot be counted with which is certainly one personal favorite proof within mathematics. One can attribute this decision to study mathematics effectively to the moment I first comprehended the evidence and its far-reaching implications (Zanini et al., 2017).

Theorem 6.5. Suppose A is a set. No surjection is present $f:A\to P(A)$. Thus $g:A\to P(A)$ provided $g(x)=\{x\}$ is injection, which shows $|A|<|P(A)|$. Particularly, P(N) is uncountable.

Proof. Consider a set A and a function $f: A \to P(A)$ to their corresponding power sets. One will demonstrate that f is not surjective (Fremlin et al., 1996). Provide a definition for a set

$$D=\{x\in A : x\in / f(x)\}.$$

It is evident that D seems a subset for A, therefore $D \in P(A)$.

Assume, for the purpose of contradiction, if f is surjective (Hemaspaandra, 1996). Therefore, it is necessary for the range of the function to include the set D. Suppose $a \in A$ be an element of A such that $f(a) = D$.

Question: If $a \in D$?

It is either true or false, without a doubt. One analyzes both scenarios.

Case 1: Yes. When $a \in D$, thus $a \in f(a)$ (since $f(a) = D$). Upon reviewing the definition for D, it becomes apparent that a cannot be included in D. This is an insurmountable task.

Case 2: No. When $a \in /$ D then $a \in / (a)$. Upon revisiting the definition of D, it becomes evident that this specifically denotes that a belongs to D. Once more, this is an unattainable task. Regardless of the scenario, a contradiction arises. Given that there are only two situations, it may be concluded that the assumption of f being surjective is untrue, so concluding the argument (Bodirsky et al., 2013).

The implications of this proof are far-reaching. Just for a taste, it implies that if you keep iterating the power set operation (Delhommé et al., 2008), you get sets with strictly larger and larger cardinalities. That is, for example:

$$|N|<|P(N)|<|P(P(N))|<|P(P(P(N)))|<\cdots$$

There exists an infinite number of different magnitudes of infinity!

5.6. RIGOROUS DEFINITIONS AND FORMALISMS

These are two definitions which one previously mentioned as being helpful in the earlier conversations, and that one will now provide clear explanations for, as one has now established the concept of countability (Osterweil, 2009).

To recapitulate part of that discourse, countability is a concept that pertains to the quality of being "small" in our context. Despite the vast R_{usual}, and intricate nature of the presence of a countable set (such as Q) that is densely distributed inside it helps to streamline some aspects. The space may be vast, but there exists a finite (i.e., small) set which is in proximity to everything (Simon, 1978).

Similarly, it is known that R_{usual} possesses numerous open sets which can become highly intricate. One has already determined that we would rather define a more straightforward set of open sets as a basis, rather than explicitly listing every open set within the topology. However, even the typical foundation of R_{usual} which involves gathering ⊠-balls round all the points, is considered large because to its uncountable nature. Nevertheless, demonstrated in both the Big List issue and Exercise 5.7 over that the collection of intervals of rational endpoints serves as a countable foundation of the standard topology. This is highly advantageous, since it reveals that the entirety of the information contained within this extensive and intricate topology may be retrieved from a minimal quantity of information (Bell & Wang, 2000).

Throughout the course, these concepts will reappear multiple times. Currently, one assign names to them.

Definition 7.1. One topological space (X, T) is deemed separable if it contains a countable and dense subset (Denecker et al., 2001).

Definition 7.2. Any topological space (X, T) is considered second measurable if there exists a countable basis upon X that produces T.

There are two important points to consider regarding second countability:

It is important to clarify that this definition is not suggesting that every basis for the topology remains countable. Rather, it states that there exists at least a single countable basis. The standard basis for the usual topology upon R is quite extensive, but the presence of B_Q demonstrates that R_{usual} was second countable (Glendening & Weinhold, 1998). Furthermore, it may appear unusual to provide a definition for "second countable" with no first establishing the definition of "first countable." Both properties have really awful names. One may have briefly mentioned the idea of being "first countable" in lecture, however not yet prepared to delve into it further. One will quickly discover that it is a nuanced and valuable characteristic, but after one has delved into sequences (Dodd & Greider, 1966).

Lastly, there is another topological property that pertains to the concept of countability. One shall not be demonstrating anything with this value right now, but you'll get some practice with it during the Big List issue in this section. One will be coming back to this belonging's multiple times all over the course (Kennedy, 2013).

Definition 7.3. A topological space (X, T) possesses the measurable chain criteria if it does not contain any uncountable collections over mutually disjoint, open subsets of X and nonempty. Another method to describe this quality is to state that any set of open subsets of X which are mutually disjoint and nonempty can be counted (Keeton & Petters, 2005).

Remark 7.4. If (X, T) a topological space (X, T) satisfies the countable chain situation, it is commonly referred to as having the ccc or being ccc.

You demonstrated that R_{usual} is compatible with a challenge presented at the conclusion of the initial segment of the Big List, however it should be noted that you were not aware of this name at that moment (Bosso, 2018).

Currently, one will include this property in our expanding collection of topological properties. However, it is worth mentioning that the connections among the property, second countability, separability, are intriguing and warrant more investigation. Whenever one establishes proof for one of those three properties, it is important to consider the other qualities within the same context (Reiter & Sørensen, 2012).

ACTIVITY 5.1.

Investigate the concept of universal countability by exploring how countability manifests in different mathematical contexts, such as in topology, analysis, and computability theory, fostering a deeper appreciation for the role of countability in diverse mathematical disciplines.

SUMMARY

Chapter 5 delves into the intricate and foundational concept of countability within the realm of topology. The chapter begins with an overview of counting, providing the essential background necessary for understanding more advanced topics. It systematically explores the foundations of countability, highlighting key principles and definitions that form the bedrock of this topic.

The chapter proceeds to discuss elementary examples and fundamental theorems, illustrating the practical applications and significance of countable sets in topology. These examples serve to ground the reader's understanding in concrete terms, paving the way for a deeper exploration of more complex topics.

Advanced examples and significant theorems are then introduced, demonstrating the breadth and depth of countability in mathematical theory. These sections emphasize the importance of countability in various topological constructs and provide insights into its role in broader mathematical contexts.

The investigation of universal countability follows, where the concept is examined through a theoretical lens. This section challenges the reader to consider the implications of countability on a universal scale, fostering a deeper appreciation for the scope and impact of this concept.

The chapter concludes with a focus on rigorous definitions and formalisms. This final section emphasizes the importance of precise language and formal approaches in the study of countability, ensuring that the reader is equipped with the tools necessary to engage with complex topological problems rigorously and effectively.

Throughout, the chapter underscores the importance of countability in topology, demonstrating its relevance through a combination of theoretical exploration and practical application. The inclusion of key terms and structured learning objectives ensures that readers can navigate the content effectively, building a solid foundation for further study in topology.

MULTIPLE CHOICE QUESTIONS

1. **Which of the following sets is countable?**
   a. The set of real numbers
   b. The set of natural numbers
   c. The set of all points on a line segment
   d. The set of all points in a plane
2. **A set is said to be countable if it can be put into a one-to-one correspondence with:**
   a. The set of real numbers

b. The set of natural numbers
c. The set of irrational numbers
d. The set of complex numbers

3. Which theorem asserts that the union of a countable number of countable sets is countable?

a. Cantor's Theorem
b. Zorn's Lemma
c. Countable Union Theorem
d. Lindelöf's Theorem

4. In topology, a space is called second-countable if it has:

a. A countable basis
b. A countable number of open sets
c. A finite basis
d. A finite number of open sets

5. Which of the following is a property of a second-countable space?

a. It is always compact
b. It is always metrizable
c. It has a countable dense subset
d. It has a finite dense subset

REVIEW QUESTIONS

1. Explain the difference between countable and uncountable sets. Provide examples of each.
2. Describe the significance of the countability axiom in topology.
3. How does the concept of countability relate to the basis of a topology?
4. What are some fundamental theorems related to countability in topology? Summarize their importance.
5. Discuss the concept of universal countability and its implications in topological spaces.

Answers to Multiple Questions

1. (b) 2. (b) 3. (c) 4. (a) 5. (c)

REFERENCES

1. Alkazwini, A. A. (2016). Noun Countability; Count Nouns and Non-count Nouns, What are the Syntactic Differences Between them?. International Journal of Applied Linguistics and English Literature, 5(7), 96–100.
2. Artin, M., & Mumford, D. (1972). Some elementary examples of unirational varieties which are not rational. Proceedings of the London Mathematical Society, 3(1), 75–95.
3. Baer, R. (1944). The fundamental theorems of elementary geometry. An axiomatic analysis. Transactions of the American Mathematical Society, 56, 94–129.
4. Bai, H., Mao, J., & Chan, S. H. G. (2022). A survey on deep learning-based single image crowd counting: Network design, loss function and supervisory signal. Neurocomputing, 508, 1–18.
5. Bakker, B., van den Bos, H., Lansdorp, P. M., & Foijer, F. (2015). How to count chromosomes in a cell: An overview of current and novel technologies. BioEssays, 37(5), 570–577.
6. Ball, K. (1997). An elementary introduction to modern convex geometry. Flavors of geometry, 31(1–58), 26.
7. Bell, D. A., & Wang, H. (2000). A formalism for relevance and its application in feature subset selection. Machine learning, 41, 175–195.
8. Bodirsky, M., Pinsker, M., & Tsankov, T. (2013). Decidability of definability. The Journal of Symbolic Logic, 78(4), 1036–1054.
9. Bogopolski, O., & Zastrow, A. (2012). The word problem for some uncountable groups given by countable words. Topology and its Applications, 159(3), 569–586.
10. Bosso, P. (2018). Rigorous Hamiltonian and Lagrangian analysis of classical and quantum theories with minimal length. Physical Review D, 97(12), 126010.
11. Cappo, M., Harvey, E., & Shortis, M. (2006, August). Counting and measuring fish with baited video techniques: an overview. Tasmania: Australian Society for fish biology. In Australian Society for Fish Biology Workshop Proceedings. Vol. 1, (pp. 101–114).
12. Charosh, M. (1944). Unifying Elementary Mathematics by Means of Fundamental Concepts. National Mathematics Magazine, 19(2), 78–90.
13. Chater, N., & Vitányi, P. M. (2003). The generalized universal law of generalization. Journal of Mathematical Psychology, 47(3), 346–369.
14. Darji, U., Elekes, M., Kalina, K., Kiss, V., & Vidnyánszky, Z. (2019). The structure of random automorphisms of countable structures. Transactions of the American Mathematical Society, 371(12), 8829–8848.
15. De Belder, M. (2011). A morphosyntactic decomposition of countability in Germanic. The journal of comparative Germanic linguistics, 14, 173–202.
16. Deal, A. R. (2016). Do all languages make countability distinctions? Evidence from Nez Perce. In Proceedings of Sinn und Bedeutung. Vol. 20, (pp. 180–197).
17. Deal, A. R. (2017). Countability distinctions and semantic variation. Natural Language Semantics, 25(2), 125–171.

18. Delbaen, F., & Schachermayer, W. (1994). A general version of the fundamental theorem of asset pricing. Mathematische annalen, 300(1), 463–520.
19. Delhommé, C., Laflamme, C., Pouzet, M., & Sauer, N. (2008). Indivisible ultrametric spaces. Topology and its Applications, 155(14), 1462–1478.
20. Denecker, M., Bruynooghe, M., & Marek, V. (2001). Logic programming revisited: Logic programs as inductive definitions. ACM Transactions on Computational Logic, 2(4), 623–654.
21. Dodd, L. R., & Greider, K. R. (1966). Rigorous solution of three-body scattering processes in the distorted-wave formalism. Physical Review, 146(3), 675.
22. Eisermann, M. (2012). The fundamental theorem of algebra made effective: an elementary real-algebraic proof via Sturm chains. The American Mathematical Monthly, 119(9), 715–752.
23. Fekete, I., Gregorics, T., Pusztai, K. K., & Veszprémi, A. (2019). Programming theorems and their applications. Teaching Mathematics and Computer Science, 17(2), 213–241.
24. Fremlin, D., Johnson, R., & Wajch, E. (1996). Countable network weight and multiplication of Borel sets. Proceedings of the American Mathematical Society, 124(9), 2897–2903.
25. Furuta, T. (2008). Concrete examples of operator monotone functions obtained by an elementary method without appealing to Lowner integral representation. Linear algebra and its applications, 429(5), 972–980.
26. Gabbay, M. J., & Mathijssen, A. (2009). Nominal (universal) algebra: equational logic with names and binding. Journal of Logic and Computation, 19(6), 1455–1508.
27. Gauffriau, M. (2021). Counting methods introduced into the bibliometric research literature 1970–2018: A review. Quantitative Science Studies, 2(3), 932–975.
28. Gieres, F. (2000). Mathematical surprises and Dirac's formalism in quantum mechanics. Reports on Progress in Physics, 63(12), 1893.
29. Glendening, E. D., & Weinhold, F. (1998). Natural resonance theory: I. General formalism. Journal of computational chemistry, 19(6), 593–609.
30. Greenberg, O. W. (2006). Why is Fundamental?. Foundations of Physics, 36(10), 1535–1553.
31. Gulgowski, P., Błaszczak, J., & Puhacheuskaya, V. (2021). The influence of aspect on the countability of Polish deverbal nominalizations: Evidence from an acceptability rating study. Zeitschrift für Sprachwissenschaft, 40(1), 1–30.
32. Heathcote, A. (2021). Countability and self-identity. European Journal for Philosophy of Science, 11(4), 109.
33. Hemaspaandra, E. (1996). The price of universality. Notre Dame Journal of Formal Logic, 37(2), 174–203.
34. Huntington, E. V. (1906). The fundamental laws of addition and multiplication in elementary algebra. Annals of Mathematics, 8(1), 1–44.

35. Kandić, M., & Taylor, M. A. (2018). Metrizability of minimal and unbounded topologies. Journal of Mathematical Analysis and Applications, 466(1), 144–159.

36. Kechris, A. S. (1975). The theory of countable analytical sets. Transactions of the American Mathematical Society, 202, 259–297.

37. Keeton, C. R., & Petters, A. O. (2005). Formalism for testing theories of gravity using lensing by compact objects:<? format?> Static, spherically symmetric case. Physical Review D—Particles, Fields, Gravitation, and Cosmology, 72(10), 104006.

38. Kennedy, J. (2013). On formalism freeness: Implementing Gödel's 1946 Princeton bicentennial lecture. Bulletin of Symbolic Logic, 19(3), 351–393.

39. Kreisel, G. (1958). Mathematical significance of consistency proofs. The Journal of Symbolic Logic, 23(2), 155–182.

40. Lachlan, A. H. (1984). On countable stable structures which are homogeneous for a finite relational language. Israel Journal of Mathematics, 49, 69–153.

41. Lecomte, D. (2020). On the complexity of Borel equivalence relations with some countability property. Transactions of the American Mathematical Society, 373(3), 1845–1883.

42. Lima, S., & Rothstein, S. (2020). A typology of the mass/count distinction in Brazil and its relevance for mass/count theories. Linguistic Variation, 20(2), 174–218.

43. Liu, L., Lu, H., Xiong, H., Xian, K., Cao, Z., & Shen, C. (2019). Counting objects by blockwise classification. IEEE Transactions on Circuits and Systems for Video Technology, 30(10), 3513–3527.

44. Lonti, Z., & Gregory, R. (2007). Accountability or countability? Performance measurement in the New Zealand public service, 1992–2002. Australian Journal of Public Administration, 66(4), 468–484.

45. Ma, J. (2008). Complete rank theorem of advanced calculus and singularities of bounded linear operators. Frontiers of Mathematics in China, 3, 305–316.

46. Makowsky, J. A. (1978). Quantifying Over Countable Sets: Positive vs Stationary Logic. Studies in Logic and the Foundations of Mathematics, 96, 183–193.

47. Mason, A. (2001). Serre's generalization of Nagao's theorem: an elementary approach. Transactions of the American Mathematical Society, 353(2), 749–767.

48. Morley, M. (1970). The number of countable models. The Journal of Symbolic Logic, 35(1), 14–18.

49. Mustafa, G. Ö. K. (2020). A mobile game experience of pre-service elementary teachers: The fundamental theorem of arithmetic. Journal of Computer and Education Research Year, 8(15), 41–74.

50. Neubrand, M. (1989). Remarks on the acceptance of proofs: The case of some recently tackled major theorems. For the Learning of mathematics, 9(3), 2–6.

51. Osterweil, L. J. (2009). Formalisms to support the definition of processes. Journal of Computer Science and Technology, 24(2), 198–211.

52. Patwal, A., Diwakar, M., Tripathi, V., & Singh, P. (2023). Crowd counting analysis using deep learning: A critical review. Procedia Computer Science, 218, 2448–2458.

53. Post, E. L. (1921). Introduction to a general theory of elementary propositions. American journal of mathematics, 43(3), 163–185.

54. Rav, Y. (1999). Why do we prove theorems?. Philosophia mathematica, 7(1), 5–41.

55. Reiter, F., & Sørensen, A. S. (2012). Effective operator formalism for open quantum systems. Physical Review A—Atomic, Molecular, and Optical Physics, 85(3), 032111.

56. Ribeiro, P., Paredes, P., Silva, M. E., Aparicio, D., & Silva, F. (2021). A survey on subgraph counting: concepts, algorithms, and applications to network motifs and graphlets. ACM Computing Surveys (CSUR), 54(2), 1–36.

57. Saghir, A., & Lin, Z. (2015). Control charts for dispersed count data: an overview. Quality and Reliability Engineering International, 31(5), 725–739.

58. Sanders, S. (2022). Countable sets versus sets that are countable in reverse mathematics. Computability, 11(1), 9–39.

59. Schmidtke, D., & Kuperman, V. (2017). Mass counts in World Englishes: A corpus linguistic study of noun countability in non-native varieties of English. Corpus Linguistics and Linguistic Theory, 13(1), 135–164.

60. Simon, B. (1978). Resonances and complex scaling: A rigorous overview. International Journal of Quantum Chemistry, 14(4), 529–542.

61. Theophilou, A. K. (1979). The energy density functional formalism for excited states. Journal of Physics C: Solid State Physics, 12(24), 5419.

62. Tsang, A. (2017). Judgement of countability and plural marking in English by native and non-native English speakers. Language Awareness, 26(4), 343–359.

63. Van Den Brom, H. E., Kerkhof, O., Lotkhov, S. V., Bogoslovsky, S. A., Willenberg, G. D., Scherer, H.,... & Piquemal, F. P. (2003). Counting electrons one by one-overview of a joint European research project. IEEE Transactions on Instrumentation and Measurement, 52(2), 584–589.

64. Wagner, D. H. (1977). Survey of measurable selection theorems. SIAM Journal on Control and Optimization, 15(5), 859–903.

65. Wang, H. (1960). Proving theorems by pattern recognition I. Communications of the ACM, 3(4), 220–234.

66. Weber, K. (2004). Traditional instruction in advanced mathematics courses: A case study of one professor's lectures and proofs in an introductory real analysis course. The Journal of Mathematical Behavior, 23(2), 115–133.

67. Wicke, H. H., & Worrell, J. M. (1976). Point-countability and compactness. Proceedings of the American Mathematical Society, 55(2), 427–431.

68. Zanini, C., Benavides-Varela, S., Lorusso, R., & Franzon, F. (2017). Mass is more: The conceiving of (un) countability and its encoding into language in 5-year-old-children. Psychonomic Bulletin & Review, 24, 1330–1340.

CHAPTER 6

Compactifications

LEARNING OBJECTIVES

After studying this chapter, you will be able to:

1. Understand the concept of compactification in topology.
2. Analyze illustrative examples of compactifications, such as (0, 1) with its usual topology.
3. Comprehend the construction and properties of one-point compactifications.
4. Apply the theorem of one-point compactification to specific examples.
5. Explore the Stone-Čech compactification and its significance in topology.
6. Differentiate between different types of compactifications and their applications.

KEY TERMS FROM THIS CHAPTER

Compactification
One-Point Compactification
Discrete Space
Topological Space
Hausdorff Space
Stone-Čech Compactification
Topological Embedding

UNIT INTRODUCTION

By now in the course, one has examined two types of spaces that are found to be highly organized and enjoyable: compact Hausdorff spaces along with metrizable spaces. These two spaces make it much easier to prove theorems compared to the usual methods. There are numerous spaces that do not fall into either category. Therefore, in general, our goal is to find a way to incorporate a given space into a space that possesses one of these desirable characteristics. Remember that a topological area (X,T) can be embedded into another (Y,U) topological space when X represents homeomorphic towards a subspace of Y (Shelton et al., 2005).

One discovered metrizability is inherited, meaning there will no benefit in stating that a space can be embedded within it. Let X is embedded toward a Y metrizable space, next homeomorphic X along with metrizable space thus metrizable its own. This can be a useful method for demonstrating that a space is metrizable (which was what one utilized when showing Urysohn's metrization theorem, for instance), but it lacks any additional insights into the properties of the space (Grana, 2006). However, it is worth noting that subspaces for compacted Hausdorff spaces may not necessarily be compact. Therefore, one might delve into concept for compact Hausdorff spaces incorporating towards topological spaces. The method referring to space is commonly known as a compactification in mathematics (Berg et al., 2011). Formalizing this concept, one will introduce a new concept known as local compactness. That explore R_{usual} typi cally lack compactness, but one will observe that it must possess certain local characteristics of compactness and be adequately structured for a compactification to be possible. Now look at the spaces R_{usual} along with Q, in instance. Neither are small, but R_{usual} has a more pleasant quality when it involves sets that occupy less space (Borel & Ji, 2006). There are many nice and simple sets that are compact within R_{usual}. However, Q has a limited number of compact sets. Open intervals such as $(a,b)\cap Q$ is not considered compact. Additionally, the closings may not be compact either. In fact, are barely containing within a single compact subsets of Q. One will provide a proof for this later in the note (Giddings et al., 2002).

There are various approaches to consider when compactifying a space. In the realm of topological spaces, there exist several intuitive methods to establish compact Hausdorff spaces that can accommodate the embedding of a given space (X,T). In this module, one will delve into two types of compactifications. Take, for instance, the interval (0, 1), which lacks compactness under its standard topology. There are multiple compact Hausdorff spaces where this space can be easily embedded. Notable examples include $[0,1],[0,1]^N$, and S^1. One will explore the various compactifications and determine their significance, distinguishing between those that are worthy of discussion and those that are unnecessary (Elitzur et al., 1991).

6.1. ILLUSTRATIVE EXAMPLES OF COMPACTIFICATIONS

In this section, one will explore two examples in a straightforward manner: The set (0, 1) via its typical topology and the set ($N,T_{discrete}$).

6.1.1. (0, 1) with its Usual Topology

There exist numerous compact Hausdorff spaces that readily accommodate the embedding of the interval (0, 1). As an illustration:

i. (0, 1) can be embedded into a subspace for [0, 1] using its natural inclusiveness map. It is worth noting that this map does not include its two endpoints, 0 and 1.

ii. The interval (0, 1) is embedded on a subspace for [0, 2] using real inclusiveness maps. 0 and all the points within the interval [1, 2] are not included in this map (Samuels,1975).

iii. The subspace (0, 1) can be embedded into S^1 using the function $f(x)=(cos(2\pi x),sin(2\pi x))$. It only fails to include the point $(1,0) in S^1$.

iv. The subspace (0, 1) can be embedded into $H=[0,1]^N$, which is known to be compact according to Tychonoff's theorem. This embedding is achieved through a map $g(x)=(x,1,1,1,\ldots)$, for instance. There are numerous approaches to accomplish this task (Arhangel'skiĭ & Wiegandt, 1975).

v. The subspace (0, 1) can be embedded within the Topologist's Sine Curve S, such as by mapping x to the point (0, x) is represented.

These examples may be excessive if one aims to represent (0, 1) like a subset for a compacted Hausdorff space. It would be more reasonable to consider simpler alternatives. It is unnecessary to assume the specified embedding towards [0, 2] since the embedding toward [0, 1] achieves similar result with no inclusion of additional distant points in "copy" of (0, 1). Embeddings through S and H exhibit even more unfavorable characteristics towards this context. It is worth noting that the images that have embeddings one had described into certain domains are densely distributed within their respective codomains. However, S^1 in the case of the embeddings to other domains, there are significant portions of their codomains that remain unaffected. In an alternative approach, the process of embedding [0, 1] into [0, 2] can be achieved by first embedding it into [0, 1] and will further embedding [0, 1] to [0, 2] using real inclusion map (Mumford, 1961).

Did you know?

Did you know that the interval (0, 1) with its usual topology is not compact? Compactification techniques can transform such spaces into compact ones, facilitating easier study and analysis.

In addition, it can be observed how the embedding to S^1appears to be the most effective or smallest one. This is because the embedding only fails to include one point, whereas embeddings toward [0, 1] fails 2 points. It demonstrates if (0, 1) get inserted in [0, 1] previously described, outcome referred to be compactification for two-point (0, 1). Embedding toward S^1 represents a one-point compactification (Johnstone, 1983).

6.1.2. N,T$_{\text{discrete}}$. Embedded in Hausdroff Space

Now turning attention to N. Embedding N toward the compact Hausdorff space is a relatively simple task. It is possible to incorporate it into a wide range of infinite compact Hausdorff spaces.

i. N is embedded like a subspace of ω + 1 through the real inclusions. One point, ω is not included in this map.

ii. The subspace N can be embedded into the $X = \frac{1}{n} : n \in N \cup 0 \subseteq R_{usual}$ each $n \rangle \rightarrow \frac{1}{n}$. Once again, this

iii. The map fails to include only one point, which is 0. X and ω + 1 were homeomorphic, as expected. In both this example and the preceding one, the image for N is densely distributed over the codomain (Nyikos, 1971).

. The N subspace for [0, 1] can be embedded as a compactification via the mapping $n \rangle\rightarrow \frac{1}{n}$. This map fails to include an infinite number of point, while resulting image might not distributed evenly throughout the codomain.

v. N is embedded into a subspace for $\omega1 + 1, via\, 1 \rangle\rightarrow m1 := min(\omega1), 2 \rangle\rightarrow m2 := min(\omega1\ \{m1\}),$, and so on. This map lacks an infinite number of points thus the image does not fill the entire codomain.

Once again, it is evident that the following pair of embeddings contains numerous unnecessary points within the codomain, whereas the first pair minimizes the addition of new points by just introducing one. In the following sections, one will establish a formal definition of a compactification and determine which compactifications were the most extensive and the most limited (García Trillos, 2020).

6.2. ONE-POINT COMPACTIFICATIONS

Definition 3.1. An embedding for (X,T) topological space into the dense subspace for a compact topological space is known as a compactification. Simply stated, a compactification refers to a specific arrangement of a topological space (Y, U) using map $f: X \to f(X)$ *acts as homeomorphism, where* $f: X \to Y$ as well $f(X) = Y$.

In analysis, one will focus solely on compactifications that exceed homeomorphism (Font & Sanchis, 2002). Therefore, won will consider two compactifications, $f_1: X \to Y_1$ and $f_2: X \to Y_2$, to be comparable when there occurs a homeomorphism $h(f_1(x)) = f_2(x)$ which preserves the embedded elements in X. In other words, $h(f_1(x)) = f_2(x)$ for all $x \in X$.

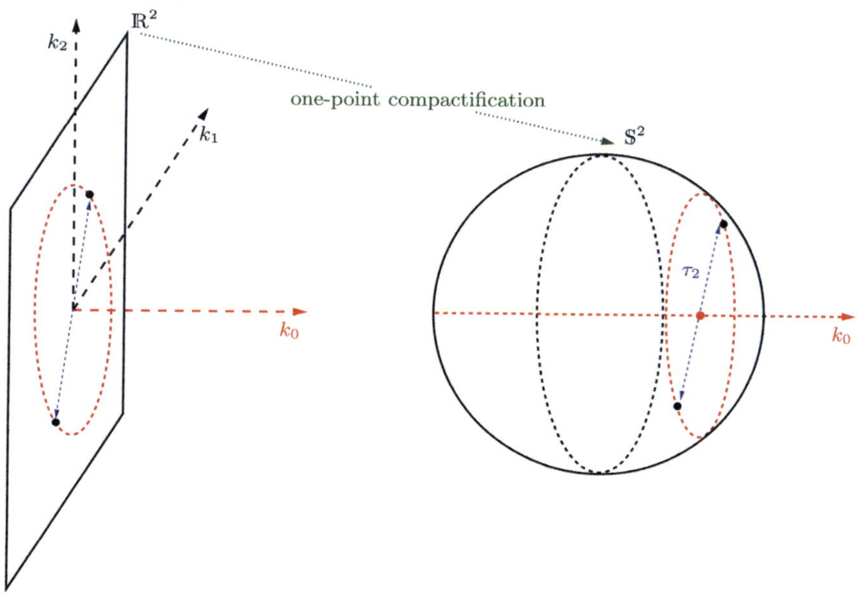

Figure 6.1. One-point compactification of the momentum spacê R d via stereographic projection and associated TRS-involution (Source: Giuseppe De Nittis, Creative Commons License)

It is important to note how this definition only considers the "efficient" examples mentioned earlier. The embeddings one defined, which contained excessive information, failed to satisfy the criteria for compactifications. They are not significant enough to warrant discussion (Roitman, 2014) (Figure 6.1).

Tip:

When studying one-point compactifications, keep in mind that adding a single point to a non-compact space can yield a compact space, exemplifying a powerful technique in topology for simplifying complex spaces.

This description does not explicitly address the concept of Hausdorffness at this point, but it will be discussed in due course. To begin, one can promptly establish a one-point compactifications that were previously discussed (Frith & Schauerte, 2017).

Definition 3.2. Consider a topological space denoted by (X, T). This one-point compactification of the given object is defined as follows:

Consider the symbol ∞ as a non-element of X, and let $X^* := X \cup \{\infty\}$. The topology T $*$ is defined on X^* by

$T^* = T \cup \{V \subseteq X^* : \infty \in V$ *and* $X \setminus V$ *is closed compact in* $X\}$.

For ease, one will symbolize the topological space (X^*, T $*$) *through* $\sigma(X)$.

6.2.1. Theorem of One-Point Compactification

Proposition 3.3. When the pair (X,T) fails to be compact, than $\sigma(X)$ serves as a compactification for $\sigma(X)$. Evidence. Let (X,T) be a non-compac topological space (Bezhanishvili & Morandi, 2011).

1. This set is compact $\sigma(X)$. For understand this, consider an open cover U for $\sigma(X)$. Where exists a subset V as an element of U which has an infinite number of elements. According to the definitions of T a compact subset*, $X^* \setminus V$ for X that is covered by U $\setminus\{V\}$. Hence, this exists a subcover (finite) for $X \setminus V$, when combine via V, forms $\sigma(X)$ finite subcover. The case might serve as a reminder of the evidence how ω +1seems compact (which is not a coincidence) (Kim & Lee, 1995).
2. The equality $\overline{X} = X^*$ holds. It is evident how $\overline{X} = X^*$ only when (X, T) seems non-compact. This can be understood quite easily, as the openness of $\{\infty\}$ in $\sigma(X)$ implies the compactness of X.
3. $i : X \to X^*$ Natural inclusion provided through embedding

$i(x)=x$. This can be readily observed, like the subsets (open) for X within σ(X) precisely correspond to components of T (Newman, 1975).

It will describes a type related compactification that is considered minimal, as it involves the addition of only one point. This method can be used to compactify any non-compact space. It is important to note, however, that initial motivation for incorporating space within compacted Hausdorff spaces, so this aspect has not been addressed thus far. Here is an instance where things did not go as planned (Rose & Hamlett, 1992).

6.2.2. Example of One-Point Compactification

Example 3.4. Let's examine Q with its customary topology. It follows that σ(Q) does not satisfy the Hausdorff property.

To demonstrate this, we will attempt to distinguish between infinity and zero using open sets. (as Q is embedded, two parts for Q may get separated by σ(Q). Let assume, purpose for contradiction, how V *and* U have open subsets for σ(Q) that do not intersect, with U containing 0 along with V containing ∞. Therefore, $U \subseteq Q \setminus V$, where V is closed or compact according to the topology upon Q∗.

However, it is not possible to achieve this. If one has an open interval for rational numbers $(a,b) \cap Q \subseteq U$ that includes 0, one can find irrational number from (a, b) along with rational numbers through (a, b) (specifically within the compact subset Q) that converge the standard topology upon R. When considering the sequence solely on Q∗, which lacks any convergent subsequences, which goes against principle which compact subsets for Q exhibit sequential compactness.

This property can be defined to prevent the occurrence of issues like the one mentioned in the example. There are numerous ways to define this concept, but one will present the simplest approach here (Beer & Vipera, 2012).

6.2.3. Topological Space and One-Point Compactification

Definition 3.5. A (X, T) topological space might consider locally compact when, for each $x \in X$, it has K compact subset for X along with U an open subset for X where $x \in U \subseteq K$. (One way to show this concept is through stating that every point within X represents a neighborhood that is small and self-contained, but I will refrain from using technical terminology in this explanation.)

Specifically, when (X, T) show Hausdorff, it will be considered locally compact when every point will an open set that having compact closure (Koushesh, 2015).

Example 3.6.

1. R_{usual} possesses local compactness and Hausdorff properties. It is already well-established the open intervals closures for closed intervals, that being compact.
2. Each discrete space has the property of being locally compact, as individual points within the space remain compact and open.
3. (X, T) a compact space can be considered compact by selecting $U=K=X$ belonging any point.

4. The Q cannot be compact locally, although it represents Hausdorff. One has already demonstrated this earlier.
5. The set σ(Q) was locally compact due to its compactness, yet it is not Hausdorff.

6. $R_{Sorgenfrey}$ is worth noting that it cannot be locally compact (but it is Hausdorff), although this can be a bit more challenging to understand. It can be deduced from the (unexpected) observation that each of the compact subsets for the Sorgenfrey line were countable.
7. Several basic features of local compactness are discussed, with the proofs being primarily leftover as tasks (Han & Na, 2018):

Proposition 3.7.

1. Local compactness remains unchanged under topological transformations.
2. Continuous pictures for spaces which locally compact might not retain the property of being locally compact.
3. Local compactness does not pass down via subspaces, but both open and closed subsets for locally compacted Hausdorff spaces possess the property of local compactness.
4. Local compactness exhibits finite productivity but not countable productivity. A product for locally compact regions defines locally compact when only a finite number of spaces remains compact (Insall et al., 2014).

Proof.

Exercise.

1. Remember each topological space can be represented as continuous mapping for a discrete space (Dydak, 2006).
2. One understands that hereditary traits are not applicable in this case, as $Q \subseteq R$. The proof over either open or closed subsets for nearby compact Hausdorff spaces represents a simple exercise.
3. One understands that hereditary traits are not applicable in this case, as $Q \subseteq R$. The proof over open as well as closed subsets on locally compacted Hausdorff spaces can be a simple exercise.
4. Proof over finite products can be a straightforward exercise. One can try proving a broader fact (it's not too challenging), but one won't be using it for the rest of this note (Mack et al., 1972).

Proposition 3.8. If (X, T) as topological space is both locally Hausdorff and compact, then the Hausdorff property also holds for the one-point compactification of X, denoted as σ(X)

Proof. (⇒). Suppose $x, y \in X^*$ be two different points. When x, y ∈ X, Given that X is a Hausdorff space, one can use open sets to divide them. For the sake of simplicity, let say that $y = \infty$. Given a locally compact X, there exists a $K \subseteq X$ compact set, and $U \subseteq X$ an open set, such as $x \in U \subseteq K$ along with X* \ K are separate open subsets for X* that do not intersect, and they separate x and y. (⇐) Assuming σ(X) shows Hausdorff. Afterwards X is a Hausdorff space because represents homeomorphic of a subspace that is a Hausdorff space. X seems locally compact because it is homeomorphic for an open subset within the σ(X) locally compacted space Here is an interesting fact: remember that in topology, a space (X, T) is considered completely typical if there exists an ongoing function which separates any point from a x ∈ X closed set that lacks that point $f: X \to [0,1]$ like $f(x)=0$ or $f(c)=1$ for every $c \in C.(X,T)$ is known as $T_{3.5}$ or Tychonoff if it wascharacterized by complete regularity and Hausdorff properties. After analyzing the Big List, it was observed that the Tychonoff property is more powerful than T_3. Urysohn's Lemma demonstrates that any T_4 space seems Tychonoff. Furthermore, it is readily apparent that the characteristic of being Tychonoff gets passed down through generations (Singha & Roy, 2024).

Proposition 3.9. It is worth noting that while each locally compacted Hausdorff space seems Tychonoff, the opposite implication does not hold in general.

Proof. Let's assume that (X, T) constitutes a locally compacted Hausdorff space. The σ(X) 1 point compactification constitutes a compact Hausdorff space that possesses the T_4 property and is also Tychonoff. Afterwards X is homeomorphic for σ(X) \ {∞}, making it Tychonoff because its trait to be Tychonoff is inherited.

On the other hand, Q is T_4 resulting in Tychonoff, but it fails to be locally compact (O'Brien et al., 1986).

6.3. THE STONE-ČECH COMPACTIFICATION

While discussing one-point compactification, it was noted that the most minimal form of compactification for a topological space. This is achieved by adding a point. One will explore the contrasting concept: the largest possible compactification for a topological space. One will explore this concept by examining few more simple findings (Budak et al., 1994).

The next result indicates that truly supports discussing spaces where a space X incorporates densely. This statement implies that if X is included as a subset from the compacted Hausdorff space Z, that there exists a compact version of X included within Z.

Proposition 5.1. Consider a topological space (X, T) along with let $h : X \rightarrow Z$ be an embedding within the Z compact Hausdorff space. Here exists a function of compactification $f : X \rightarrow Y$ for X in which there exists an embedding function $g : Y \rightarrow Z$, satisfying the requirement that $h(x) = g(f(x))$ for all $x \in X$. Furthermore, each Y compactification is uniquely determined for equivalence (Walker, 1974).

Remark 5.2. This result may appear intricate, but it is in fact very simple. The statement regarding the theorem is most effectively demonstrated with a diagram, that is a visual representation similar to the one shown below:

The map h throughout the top reflects the provided embedding for X through a compact Hausdorff space. The theorem states that there is both a compactification f : $X \rightarrow Y$ along with an embedding g of Y in Z.The last requirement in the theorem states that the diagram is commutative. This means that if one travel from X to Z using either of the two sets of arrows, the resultant function compositions will be identical.

Proof:

Consider $(h : X \rightarrow Z)$ Bieng embedded in a space that is both compact and Hausdorff.

Thus $(h : X \rightarrow h(X))$ shows homeomorphism. Suppose $\left(Y = \overline{h(X)} \subseteq Z\right)$. Now (Y) within a closed subset within compact Hausdorff space which means it serves as a compactification in (X) through the map (h). It is evident that (Y) is embedded into (Z) through a natural inclusion map $(i : Y \rightarrow Z)$, that possesses the necessary properties (Curi, 2010).

Proving the distinctive characteristics of a compactification (Y) beyond equivalence can be quite intriguing and challenging. Assume $(f: X \to Y')$ is another way to represent (X) that can be included in (Z) through a map $(g: Y' \to Z)$ like $(h(x) = g(f(x)))$ Thus $(g(f(X)) = \overline{h(X)})$

It is demonstrated that the $(g(Y') = \overline{h(X)})$. Firstly, it's worth noting that (g) is a continuous function. Additionally, $(Y' = \overline{f(X)})$ due to its compactification nature. Thus, $(g(Y') = g(\overline{f(X)}) \subseteq \overline{h(X)}$ \). On the other hand, $(g(Y'))$ constitutes the continuous image belonging to a compact set, making it compact and, as a result, closed (Z) is Hausdorff. Similarly, $(\overline{h(X)} = g(f(X)) \subseteq g(Y'))$, thus $(\overline{h(X)} \subseteq g(Y'))$ since $(g(Y')$ \) remains closed.

After establishing this, it becomes clear that a map $(g^{-1} \circ i: Y \to Y')$ serves as a homeomorphism that has the desired property. That is, one demonstrated $(Y = g(Y'))$ and given that (g) represents an embedding, with not much else to be done (Curi, 2007).

Once more, this result simply states that when a space (X, T) is embedded toward the compacted Hausdorff space Z, it is necessary to go by a compactification in the process. One is essentially incorporating X through 1 of their compactifications, while incorporating the compactification to Z.

It seems unlikely that an intriguing "largest" Y compactification exists and consistently serves an intermediary compactification, that was stated earlier. Space might be included in its one-point compactification, meaning that Y of this nature cannot be larger than that.

One could achieve significant success, nonetheless. In this a Tychnoff space X exists a distinct (yet equivalent) compactification that possesses a certain understanding of every continuous mappings through X to any compact Hausdorff spaces. This is known as the Stone-Čech compactification. There is a significant amount of theory surrounding this design, which can be carried out with different levels of generality. One provides a comprehensive analysis including a focused examination of the case involving natural numbers (Šobot, 2015).

6.3.1. Stone-Čech Compactification of a Discrete Space

While working in the Stone-Čech compactification for a discrete topological space, there is a more elegant way to describe it using ultrafilters (Glicksberg, 1959). One will discuss the scenario of N that has a discrete topology, although the concepts one present can be applied in any distinct space. Characterizing the Stone-Čech compactification in a $T_{3.5}$ space in regard to ultrafilters can be quite challenging, especially in non-discrete spaces (Porter & Woods, 1972).

Remember that the ultrafilter U upon N is considered principal when either of its two similar properties holds true (Keesling, 1977).

i. U is a finite set.

ii. There represent an $n \in N$ like $U = Un := \{A \subseteq N : n \in A\}$.

A non-principal ultrafilter is one which is not principal. Alternatively, U is considered non-principal when, for every natural number n, the set $n \in N, \{n, n+1, n+2, \ldots\} \in U$. (Be sure to thoroughly understand this equivalence) (Brandal & Brandal, 1979).

ACTIVITY 6.1.

Explore the Stone-Čech compactification by considering how it extends a given topological space to include all continuous functions from that space to the real numbers, illustrating its broad applicability in analysis and functional analysis.

SUMMARY

Chapter 6 of this textbook delves into the intricate concept of compactifications within the realm of topology. It begins by elucidating the notion of compactification through illustrative examples, notably exploring how the interval (0, 1) behaves under its usual topology. The discussion extends to embedding discrete spaces like (N) with the discrete topology into Hausdorff spaces, showcasing fundamental applications of compactification in practical scenarios.

A focal point of the chapter is the exploration of one-point compactifications, where the theoretical framework is established through theorems and concrete examples. This section highlights how adding a single point can render a non-compact space into a compact one, illustrating its utility in various topological contexts. Furthermore, the Stone-Čech compactification is introduced as a general method to construct compact spaces from arbitrary topological spaces, emphasizing its role in extending the realm of compact topology.

Throughout, the chapter emphasizes rigorous definitions and formalisms, providing readers with a deep understanding of compactifications and their applications. It concludes with a concise summary that encapsulates the key concepts discussed, reinforcing the importance of compactifications in advancing the study and application of topology.

MULTIPLE CHOICE QUESTIONS

1. Which of the following is a key feature of compactification?
a. Adding isolated points
b. Removing points to form a smaller space
c. Extending a space to include its limits
d. Shrinking the space to a single point
2. The one-point compactification of a space (X) adds which type of point?
a. An arbitrary point
b. A point at infinity
c. A point at the origin
d. A point with zero coordinate
3. What role does the Stone-Čech compactification play in topology?
a. It compactifies any topological space
b. It only compactifies discrete spaces
c. It compactifies metric spaces
d. It compactifies only connected spaces
4. When can a non-compact space be compactified using the one-point method?
a. When it has infinitely many points

b. When it has a dense set

c. When it has an isolated point

d. When it has an open covering

REVIEW QUESTIONS

1. Explain the concept of compactification with examples.
2. How does one-point compactification transform a non-compact space into a compact one?
3. What is the significance of the Stone-Čech compactification in topology?
4. Compare and contrast the one-point compactification and the Stone-Čech compactification.
5. Discuss the conditions under which a space can be compactified using the one-point method.
6. Why is the interval (0, 1) under its usual topology a significant example in compactification?
7. How does the choice of topology influence the compactification process?

Answers to Multiple Questions

1. (c) 2. (b) 3. (a) 4. (b)

REFERENCES

1. Arhangel'skiĭ, A. V., & Wiegandt, R. (1975). Connectednesses and disconnectednesses in topology. General Topology and its Applications, 5(1), 9–33.
2. Beer, G., & Vipera, M. C. (2012). The Alexandroff one-point compactification as a prototype for extensions. Advances in Mathematics, 231(3–4), 1598–1618.
3. Berg, M., Marsh, D., McAllister, L., & Pajer, E. (2011). Sequestering in string compactifications. Journal of High Energy Physics, 2011(6), 1–44.
4. Bezhanishvili, G., & Morandi, P. J. (2011). One-point order-compactifications. Houston J. Math, 37(3), 699–713.
5. Borel, A., & Ji, L. (2006). Compactifications of symmetric and locally symmetric spaces. In Lie Theory: Unitary Representations and Compactifications of Symmetric Spaces. Boston, MA: Birkhäuser Boston. (Vol. 3, pp. 69–137).
6. Brandal, W., & Brandal, W. (1979). The Stone-Čech compactification of N. Commutative Rings whose Finitely Generated Modules Decompose, 2(1), 49–57.
7. Budak, T., Işik, N., & Pym, J. (1994, July). Subsemigroups of Stone-Čech compactifications. In Mathematical Proceedings of the Cambridge Philosophical Society. Cambridge University Press. (Vol. 116, No. 1, pp. 99–118).

8. Curi, G. (2007). Exact approximations to Stone–Čech compactification. Annals of Pure and Applied Logic, 146(2–3), 103–123.

9. Curi, G. (2010). On the existence of Stone-Čech compactification. The Journal of Symbolic Logic, 75(4), 1137–1146.

10. Dydak, J. (2006, October). Local n-connectivity of quotient spaces and one-point compactifications. In Shape Theory and Geometric Topology: Proceedings of a Conference Held at the Inter-University Centre of Postgraduate Studies, Dubrovnik, Yugoslavia, January 19–30, 1981. Berlin, Heidelberg: Springer Berlin Heidelberg. Berlin, Heidelberg: Springer Berlin Heidelberg. (Vol. 66, pp. 48–72).

11. Elitzur, S., Forge, A., & Rabinovici, E. (1991). Some global aspects of string compactifications. Nuclear Physics B, 359(2–3), 581–610.

12. Font, J. J., & Sanchis, M. (2002). A characterization of locally compact spaces with homeomorphic one-point compactifications. Topology and its Applications, 121(1–2), 91–104.

13. Frith, J., & Schauerte, A. (2017). One-point compactifications and continuity for partial frames. Categories and General Algebraic Structures with Applications, 7(Special Issue on the Occasion of Banaschewski's 90th Birthday 2(3), 57–88.

14. García Trillos, N. (2020). Gromov–Hausdorff limit of Wasserstein spaces on point clouds. Calculus of Variations and Partial Differential Equations, 59(2), 73.

15. Giddings, S. B., Kachru, S., & Polchinski, J. (2002). Hierarchies from fluxes in string compactifications. Physical Review D, 66(10), 106006.

16. Glicksberg, I. (1959). Stone-Čech compactifications of products. Transactions of the American Mathematical Society, 2(2), 369–382.

17. Grana, M. (2006). Flux compactifications in string theory: A comprehensive review. Physics reports, 423(3), 91–158.

18. Han, S. E., & Na, I. K. (2018). Topologies associated with the one point compactifications of Khalimsky topological spaces. Topology and its Applications, 241, 333–344.

19. Insall, M., Loeb, P. A., & Marciniak, M. A. (2014). End compactifications and general compactifications. Journal of Logic and Analysis, 6, 651–655.

20. Johnstone, P. T. (1983). The point of pointless topology. Bulletin of the American Mathematical Society, 8(1), 41–53.

21. Keesling, J. (1977). The Stone-Čech compactification and shape dimension. In Topology Proceedings (Vol. 2, pp. 483–508).

22. Kim, H. J., & Lee, K. B. (1995). One Point Compactification in Semiflows. Journal of the Chungcheong Mathematical Society, 8(1), 167–171.

23. Koushesh, M. R. (2015). One-point connectifications. Journal of the Australian mathematical society, 99(1), 76–84.

24. Mack, J., Rayburn, M., & Woods, G. (1972). Local topological properties and one point extensions. Canadian Journal of Mathematics, 24(2), 338–348.

25. Mumford, D. (1961). The topology of normal singularities of an algebraic surface and a criterion for simplicity. Publications Mathématiques de l'IHÉS, 9, 5–22.

26. Newman, S. (1975). One-point compactifications of Q-manifold factors and infinite mapping cylinders. In Studies in Topology. Academic Press. (Vol. 3, pp. 411–426).

27. Nyikos, P. (1971). Not every 0-dimensional realcompact space is N-compact. Bulletin of the American Mathematical Society, 77(3), 392–396.

28. O'Brien, T., Rajagopalan, M., & Satyanarayana, M. (1986, December). Semigroups and their one-point compactifications. In Semigroup forum. Springer-Verlag. (Vol. 33, pp. 391–404).

29. Porter, J. R., & Woods, R. G. (1972). Nowhere dense subsets of metric spaces with applications to Stone-Čech compactifications. Canadian Journal of Mathematics, 24(4), 622–630.

30. Roitman, J. (2014). Box products of one point compactifications and related results. In Topology Proceedings (Vol. 44, pp. 197–206).

31. Rose, D. A., & Hamlett, T. R. (1992). On one-point $\cal I $-compactification and local $\cal I $-compactness. Mathematica Slovaca, 42(3), 359–369.

32. Samuels, P. (1975). A topology formed from a given topology and ideal. Journal of the London Mathematical Society, 2(4), 409–416.

33. Shelton, J., Taylor, W., & Wecht, B. (2005). Nongeometric flux compactifications. Journal of High Energy Physics, 2005(10), 085.

34. Singha, M., & Roy, S. (2024). Influence of ideals in compactifications. Mathematica Slovaca, 74(3), 737–746.

35. Šobot, B. (2015). Divisibility in the Stone-Čech compactification. Reports on Mathematical Logic, (50), 53–66.

36. Walker, R. C. (1974). Development of the Stone-Čech Compactification. In The Stone-Čech Compactification. Berlin, Heidelberg: Springer Berlin Heidelberg. (Vol, 2, pp. 1–42).

CHAPTER

7 Homotopy and Fundamental Group

LEARNING OBJECTIVES

After studying this chapter, you will be able to:

1. Understand the concept of homotopy and its significance in topological spaces.
2. Describe the fundamental group and its role in classifying topological spaces.
3. Calculate fundamental groups of simple spaces using various methods.
4. Explain the properties and functoriality of the fundamental group.
5. Analyze the relationship between fundamental groups and covering spaces.
6. Apply algebraic topology concepts such as Van Kampen's Theorem and higher homotopy groups.

KEY TERMS FROM THIS CHAPTER

Homotopy

Homotopy Invariance

Functoriality

Van Kampen's Theorem

Homotopy Equivalence

Fundamental Group

Covering Spaces

UNIT INTRODUCTION

In topology, homotopy theory and the fundamental group are foundational concepts that illuminate the fundamental structure and connectivity of topological spaces. Homotopy theory explores the notion of continuous deformations between maps, focusing on transformations that preserve essential topological properties such as connectivity and dimensionality. Homotopy equivalence between spaces indicates their topological indistinguishability under continuous transformations, providing a robust framework to study the space of all possible continuous mappings between two spaces (Adhikari, 2023) (Figure 7.1).

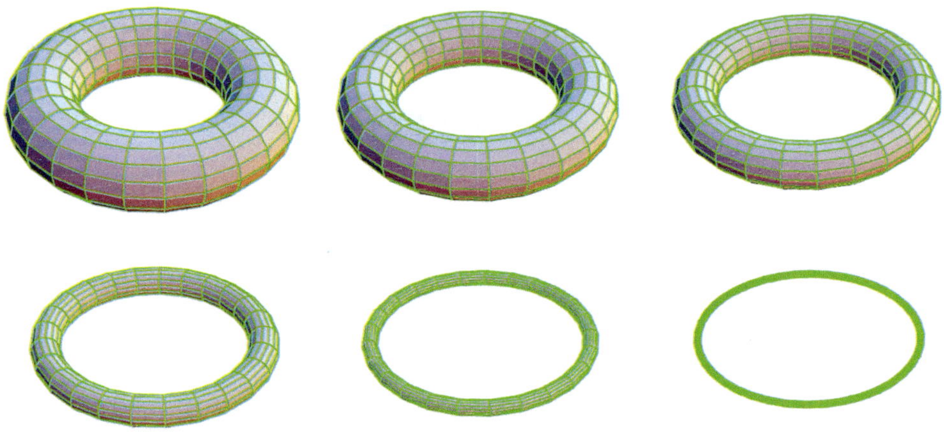

Figure 7.1. Illustration of the homotopic objects (Source: Math World, Creative Commons License)

The fundamental group, a key invariant in algebraic topology derived from homotopy theory, captures essential information about the loops and their deformations around a base point in a space. It characterizes the space's non-contractible loops and measures its topological non-triviality (Dyer, 1978). Beyond classification, the fundamental group plays a crucial role in various applications, including the study of covering spaces, group actions on spaces, and the classification of surfaces and manifolds. Together, homotopy theory and the fundamental group form the backbone of algebraic topology, offering powerful tools to analyze, classify, and understand the intricate topological properties of spaces in both pure mathematics and its applications across diverse scientific disciplines (dos Santos, 2015).

7.1. HOMOTOPY OF PATHS

Homotopy is a fundamental concept in topology that describes a continuous transformation from one path to another within a given space. Formally, let (X) be a topological space, and let $(f,g:[0,1]\to X)$ be two continuous functions (paths) with the same initial and terminal points, i.e., $(f(0)=g(0))$ and $(f(1)=g(1))$. A homotopy between (f) and (g) is a continuous function $(H:[0,1]\times[0,1]\to X)$ such that:

$$(H(s,0)=f(s))\,for\,all\,(s\in[0,1]),$$

$$(H(s,1)=g(s))\,for\,all\,(s\in[0,1]),$$

$$(H(0,t)=f(0)=g(0))\,for\,all\,(t\in[0,1]),$$

$$(H(1,t)=f(1)=g(1))\,for\,all\,(t\in[0,1]).$$

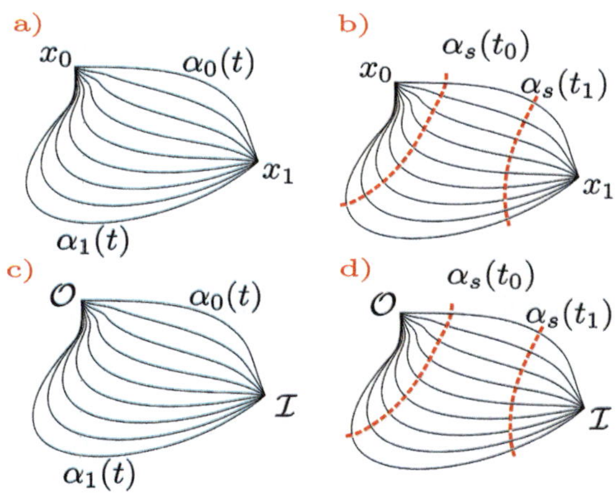

Figure 7.2. Homotopy between two paths. (a) The family of homotopic paths ${\alpha }_{s}(t)=H(t,s)$ is displayed. Points x0 = α(0) and x1 = α(1) are to are the extremes of the family of paths αs(t). Two extreme paths α0; α1, both belong of the family of paths αs (t). (b) The dashed lines is the position of family of paths at t0 and t1, where ${t}_{0},{t}_{1}\in [0,1]$. (c) For an stigmatic imaging system the homotopy ${\alpha }_{s}(t)=H(t,s)$ is the function between the function that maps the trajectories of the rays. Two extremes are the object and the images points. (d) The dashed lines are the wavefront at times t0 and t1.

This definition ensures that (H) provides a continuous deformation from (f) to (g), maintaining the endpoints fixed throughout the deformation process (Li & Yorke, 1980) (Figure 7.2).

7.1.1. Examples of Homotopic and Non-Homotopic Paths

Consider the space (R^2).Let (f) be a path that follows a straight line from point (A) to point (B), and let (g) be another path that forms an arc from (A) to (B).. These two paths can be continuously deformed into each other by gradually bending the straight line into the arc or vice versa, making (f) and (g) homotopic.

Remember

Remember that homotopy of paths explores continuous deformations between paths in a space, crucial for understanding topological equivalence and connectedness.

In contrast, consider (X) as the punctured plane $\left(R^2 \setminus \{(0,0)\}\right)$. Let (f) be a path that loops around the origin once, and let (g) be a path that does not loop around the origin. There is no continuous deformation in this space that transforms (f) into (g) without passing through the origin, so (f) and (g) are not homotopic.

7.1.2. Homotopy Equivalence and Deformation

Two continuous functions (or paths) $(f, g:[0,1] \to X)$ are said to be homotopically equivalent, denoted $(f \simeq g)$, if there exists a homotopy (H) between them (Maleki et al., 2014). Homotopy equivalence is an equivalence relation, meaning it is reflexive $((f \simeq f))$, symmetric $(if\,(f \simeq g)$, then $(g \simeq f)$ and transitive $(if\,(f \simeq g)$ and $(g \simeq h)$, then $(f \simeq h))$.

Deformation refers to the process of continuously transforming one path into another. In homotopy theory, this deformation respects the topological structure of the space, allowing for the study of the space's properties by examining the behavior of paths under continuous transformations. Homotopy equivalence is a powerful tool in algebraic topology, enabling the classification of spaces and paths based on their topological properties rather than their exact geometric forms (Hauschild, 1993).

7.1.3. Homotopy in Topological Spaces

Following is the explanation of homotopy in topological spaces.

7.1.3.1. Continuous Maps and Homotopy

In the broader context of topological spaces, homotopy extends beyond paths to continuous maps between spaces. Let (X) and (Y) be topological spaces, and let $(f, g : X \to Y)$ be continuous maps. A homotopy between (f) and (g) is a continuous map $(H : X \times [0,1] \to Y)$ such that:

1. $(H(x,0) = f(x))\, for\, all\, (x \in X)$,
2. $(H(x,1) = g(x))\, for\, all\, (x \in X)$.

This homotopy (H) provides a continuous deformation of the map (f) into the map (g). If such a homotopy exists, (f) and (g) are said to be homotopic, denoted $(f \simeq g)$. This concept allows for a flexible comparison of topological spaces, focusing on their structural properties rather than specific details (Rieser, 2021) (Figure 7.3).

Did you know?

Did you know that two paths in a space are homotopic if one can be continuously deformed into the other? This concept is fundamental in determining whether paths are topologically equivalent.

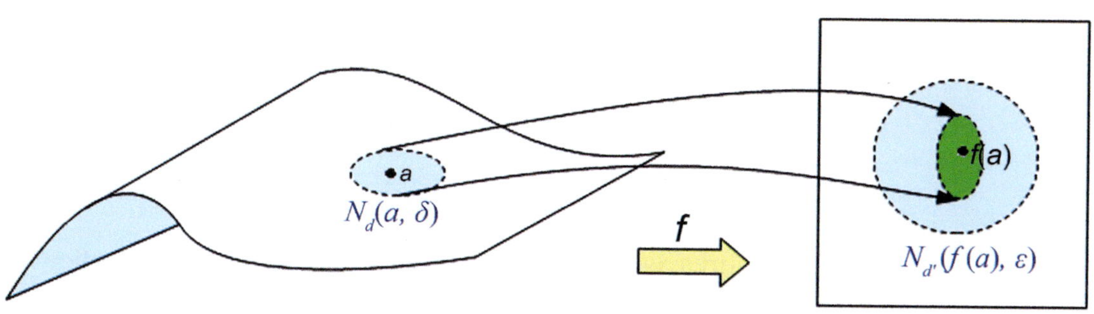

Figure 7.3. Illustration of the continuous map (Source: Math Strek, Creative Commons License)

7.1.3.2. Homotopy Relative to a Subspace

Tip:

The fundamental group captures essential information about the topological structure of a space through its equivalence classes of loops, providing a powerful tool for distinguishing spaces based on their connectivity properties.

Homotopy can also be defined relative to a subspace. Let (A) be a subspace of (X) Two maps $(f,g:X\rightarrow Y)$ are homotopic relative to (A) if there is a homotopy $(H:X\times[0,1]\rightarrow Y)$ such that:

$$1.(H(x,0)=f(x))\,and\,(H(x,1)=g(x))\,for\,all\,(x\in X),$$

$$2.(H(a,t)=f(a)=g(a))\,for\,all\,(a\in A)\,and\,(t\in[0,1]).$$

This means that (f) and (g) can be continuously deformed into each other while keeping their behavior fixed on the subspace (A). Homotopy relative to a subspace is a key tool in algebraic topology, particularly in the study of relative homotopy groups and the extension of homotopies.

7.1.3.3. Homotopy Classes of Maps

The set of all continuous maps from (X) to (Y) can be partitioned into equivalence classes under the relation of homotopy. Each equivalence class, called a homotopy class, consists of maps that can be continuously deformed into each other. The collection of these homotopy classes is denoted by $([X,Y])$. This set of homotopy classes is a fundamental object of study in algebraic topology, providing a way to classify maps between spaces based on their topological properties. The concept of homotopy classes enables a more abstract and structural approach to understanding the relationships between topological spaces (Thomas, 1964).

7.1.4. Homotopy Invariance

Following is the explanation of homotopy invariance.

7.1.4.1. Invariance Under Continuous Deformations

Homotopy invariance is a crucial property in topology, indicating that certain topological properties remain unchanged under continuous deformations. Specifically, if two maps $(f,g:X\rightarrow Y)$ are homotopic, then any topological property that is homotopy-invariant will be the same for both (f) and (g) (Marsden & Scheurle, 1987). For example, if a space is contractible, meaning it can be continuously shrunk to a point within itself, this property is preserved under homotopy. This invariance provides a robust framework for studying spaces up to homotopy equivalence, allowing topologists to classify spaces and maps based on their essential structural features rather than their precise forms (Cedzich et al., 2018).

7.1.4.2. Homotopy and Homeomorphisms

Homotopy is closely related to the concept of homeomorphisms, which are bijective continuous maps with continuous inverses. While homeomorphisms preserve the exact topological structure of a space, homotopy captures a more flexible notion of equivalence.

If two spaces (X) and (Y) are homeomorphic, they are also homotopy equivalent, as there exist continuous maps between them that can be deformed into each other (Conner & Raymond, 1977). However, the converse is not true; two spaces can be homotopy equivalent without being homeomorphic. Homotopy equivalence thus serves as a broader, more inclusive notion of equivalence in topology, allowing for the study of spaces that are similar in a more relaxed sense. This makes homotopy a powerful tool in both theoretical and applied topology (Kruggel, 1998).

7.2. FUNDAMENTAL GROUP

The fundamental group is an algebraic structure that captures essential information about the shape of a topological space. Given a topological space (X) and a point $(x_0 \in X)$ (called the base point), the fundamental group of (X) at (x_0), denoted by $(\pi_1(X, x_0))$, consists of all loop paths starting and ending at (x_0), up to homotopy equivalence. Formally, a loop is a continuous map $(\gamma:[0,1] \to X)$ such that $(\gamma(0) = \gamma(1) = x_0)$. Two loops (γ) and (δ) are considered homotopic if there exists a continuous map $(H:[0,1]\times[0,1] \to X)$ such that:

1. $(H(s,0) = \gamma(s))\, for\, all (s \in [0,1])$,

2. $(H(s,1) = \delta(s))\, for\, all (s \in [0,1])$,

3. $(H(0,t) = H(1,t) = x_0)\, for\, all (t \in [0,1])$.

The set of homotopy classes of loops forms a group under the operation of concatenation of paths, where the identity element is the constant path at (x_0) and the inverse of a path is the same path traversed in the reverse direction (Patera et al., 1975).

7.2.1. Fundamental Group of Simple Spaces

The fundamental group can vary significantly depending on the topological space considered. For simple spaces, such as the plane (R^2), any loop can be continuously contracted to a point, so the fundamental group $(\pi_1(R^2, x_0))$ is trivial, consisting only of the identity element (Ahiezer et al., 1974).

For the circle (S^1), the situation is different. Any loop on the circle can be classified by the number of times it winds around the circle, either in the clockwise or counterclockwise direction. The fundamental group $\left(\pi_1\left(S^1, x_0\right)\right)$ is isomorphic to the integer group (Z), where each integer represents a different winding number (Gorbatsevich, 1986).

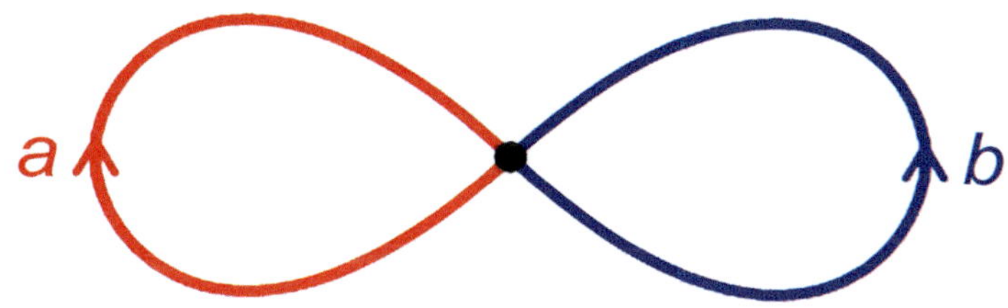

Figure 7.4. The fundamental group of the figure eight is the free group on two generators a and b (Source: Jim Belk, Creative Commons License)

7.2.2. Calculation of Fundamental Groups

Calculating the fundamental group involves finding all possible loops based at (x_0) and determining their homotopy classes. For example, to find $\left(\pi_1\left(S^1, x_0\right)\right)$, consider a loop (γ) that winds (n) times around the circle. By concatenating paths and examining their homotopy classes, we see that these loops can be associated with integers representing the winding number, thus establishing the isomorphism with (Z).

For more complex spaces, such as the torus $\left(T^2\right)$, the fundamental group calculation becomes more intricate. The torus can be represented as a product of two circles, $\left(S^1 \times S^1\right)$, leading to the fundamental group $\left(\pi_1\left(T^2, x_0\right)\right)$ being isomorphic to $(Z \times Z)$, reflecting the independent winding around each circle (Boggess & Raich, 2009).

Understanding the fundamental group provides deep insights into the topological structure of spaces, revealing how they can be deformed and manipulated while preserving essential features. This makes it a powerful tool in algebraic topology, with applications in various fields such as physics, geometry, and more (Krivoshei, 1968).

7.2.3. Properties of the Fundamental Group

Following are the properties of fundamental group.

7.2.3.1. Group Structure and Operations

The fundamental group $\left(\pi_1\left(X, x_0\right)\right)$ is endowed with a group structure, where the group operation is concatenation of loops. Given two loops (γ_1) and (γ_2) based at (x_0), their concatenation $(\gamma_1 * \gamma_2)$ is a loop that first follows (γ_1) and then (γ_2). The identity element in $(\pi_1(X, x_0))$ is the constant loop at (x_0), and the inverse of a loop (γ) is the loop traversed in the reverse direction. This group structure encapsulates the algebraic properties of loops in the space and their equivalence under homotopy, making $(\pi_1(X, x_0))$ a powerful invariant in topology (Porter, 1987).

7.2.3.2. Path Lifting and Fundamental Group

Path lifting is a crucial concept when studying the fundamental group, particularly in the context of covering spaces. Given a covering space $(p : \tilde{X} \to X)$ and a path $(\ :[0,1] \to \ )$ starting at \(x_0\), path lifting ensures that for any point $(\widetilde{x_0}) in (\tilde{X})$ such that $(p(\widetilde{x_0}) = x_0)$, there exists a unique path $(\tilde{\gamma} : [0,1] \to \tilde{X})$ starting at $(\widetilde{x_0})$ such that $(p(\tilde{\gamma}(t)) = \gamma(t))$ for all $(t \in [0,1])$. This property links the fundamental group of $(\ )$ to the fundamental group of its covering space $(\tilde{X})$, allowing us to analyze and understand the structure of more complex spaces by studying their simpler covering spaces (Biagi & Bonfiglioli, 2017).

7.2.3.3. Homotopy Lifting Property

The homotopy lifting property is another significant feature of covering spaces in relation to the fundamental group. It states that if $(F : X \times [0,1] \to Y)$ is a homotopy between two continuous maps $(f, g : X \to Y)$, and if there exists a map $(\tilde{f} : X \to \tilde{Y})$ such that $(p \circ \tilde{f} = f)$,. then there exists a homotopy $(\tilde{F} : X \times [0,1] \to \tilde{Y}) \text{lifting} (F)$ such that $(\tilde{F}(x,0) = \tilde{f}(x))$ $\text{forall}(x \in X)$ $\text{and}(p \circ \tilde{F} = F)$. This property ensures that homotopies in the base space (Y). can be lifted to the covering space $(\tilde{Y})$, preserving the structure of fundamental groups and allowing for more detailed analysis of topological spaces and their mappings (Hardeman, 2018).

7.2.4. Functoriality of the Fundamental Group

The functoriality of the fundamental group is a fundamental concept in algebraic topology, linking the topological properties of spaces to the algebraic properties of groups. Functoriality asserts that for any continuous map $(f: X \to Y)$ between topological spaces and a chosen base point$(x_0 \in X)$, there exists an induced homomorphism $(f_*: \pi_1(X, x_0) \to \pi_1(Y, f(x_0)))$ between their respective fundamental groups. This homomorphism preserves the group operation and identity, reflecting how maps between spaces interact with loops and their homotopy classes.

To define (f_*), consider a loop $(\gamma: [0,1] \to X)$ based at (x_0). The map $(f \circ \gamma: [0,1] \to Y)$ is a loop based at $(f(x_0))$, and $(f_*([\gamma]) = [f \circ \gamma])$, where $([\gamma])$ denotes the homotopy class of (γ) in $({}_1(X, x_0))$.. Functoriality ensures that (f_*) respects composition of loops and their homotopy classes, establishing a connection between topological spaces and their fundamental groups that is functorial in nature (Nori 1976).

7.2.5. Fundamental Group and Covering Spaces

Covering spaces play a crucial role in the study of the fundamental group, providing a mechanism to relate the fundamental group of a space (X) to that of its covering spaces. A covering space $(p: \tilde{X} \to X)$ induces a natural injective homomorphism $(p_*: \pi_1(\tilde{X}, \widetilde{x_0}) \to \pi_1(X, x_0))$,where$(\widetilde{x_0})$ is a chosen base point in $(\tilde{X})$ such that $(p(\widetilde{x_0}) = x_0)$.

This homomorphism (p_*) reflects how loops in (X) can be lifted to loops in $(\tilde{X})$, and vice versa, while preserving their homotopy classes. Specifically, if \$(\gamma: [0,1] \to X)$ is a loop based at (x_0),itslift$(\tilde{\gamma}: [0,1] \to \tilde{X})$ based at $(\widetilde{x_0})$ satisfies $(p \circ \tilde{\gamma} = \gamma)$.Consequently, (p_*) enables us to understand the fundamental group of (X) in terms of the fundamental group of its covering spaces, providing insights into the topological structure and connectivity of (X) through its covering space constructions (Vakil & Wickelgren, 2011).

7.3. COVERING SPACES

Following is the explanation of covering spaces.

7.3.1. Definition and Basic Properties

A covering space $\left(p:\tilde{X} \to X\right)$ is a fundamental concept in algebraic topology that relates to the notion of local triviality in the context of continuous mappings between topological spaces. Formally, a covering space $\left(p:\tilde{X} \to X\right)$ satisfies the following properties:

i. Local Homeomorphism: For every point $(x \in X)$, there exists an open neighborhood $(U \subseteq X)$ of (x) such that $\left(p^{-1}(U)\right)$ is a disjoint union of open sets in $\left(\tilde{X}\right)$, each of which is homeomorphic to (U) under (p).

ii. Covering Map: (p) is a continuous surjective map that locally resembles a homeomorphism, preserving the local structure of (X) in $\left(\tilde{X}\right)$.

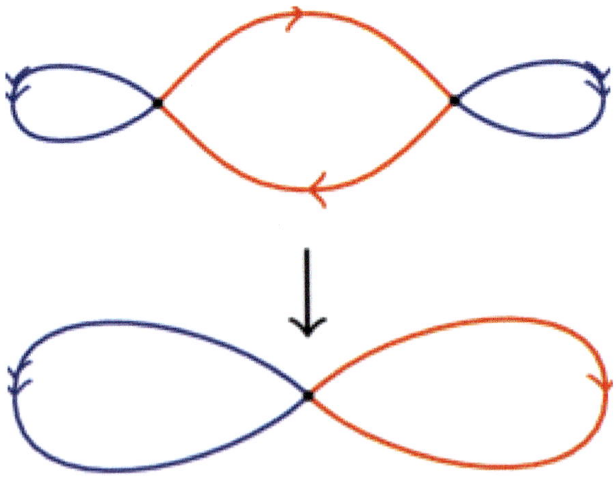

Figure 7.5. The images defines a 2-to-1 covering space (Source: Jonny Evans, Creative Commons License)

Covering spaces allow for the exploration of topological properties of (X) by lifting information from (X) to $\left(\tilde{X}\right)$ while retaining essential topological characteristics. They serve as a bridge between the local properties of (X) and the global topological features that may not be readily apparent in (X) alone (Groemer, 1986) (Figure 7.5).

7.3.2. Lifting Properties

The lifting properties of covering spaces provide essential tools for understanding how paths and mappings behave between (X) and $(\tilde{X})$. Specifically (Mironescu & Van Schaftingen, 2021):

i. Path Lifting: Given a path $(\gamma:[0,1]\rightarrow X)$ and a point $(\widetilde{x_0}\in p^{-1}(\gamma(0)))$, there exists a unique lift $(\tilde{\gamma}:[0,1]\rightarrow \tilde{X})$ such that $(\tilde{\gamma}(0)=\widetilde{x_0})$ and $(p\circ\tilde{\gamma}=\gamma)$. This property allows paths in (X) to be studied through their lifts in $(\tilde{X})$, providing insights into the connectivity and structure of (X) through the lens of $(\tilde{X})$.

ii. Covering Homotopy: A homotopy $(F:X\times[0,1]\rightarrow X)$ induces a lifting $(\tilde{F}:X\times[0,1]\rightarrow\tilde{X}\backslash)$ such that $(\tilde{F}(x,0))$ corresponds to the lift of $(x)in(\tilde{X})$ and $(p\circ\tilde{F}=F)$. This lifting property ensures that homotopies in (X) can be tracked and analyzed through their lifts in $(\tilde{X})$, maintaining the topological integrity of (X) within the covering space $(\tilde{X})$.

7.3.3. Classification of Covering Spaces

Covering spaces can be classified based on their fundamental group action on the fiber $(p^{-1}(x))$ for a chosen base point $(x\in X)$:

i. Regular Coverings: A covering $(p:\tilde{X}\rightarrow X)$ is regular if its fundamental group $(\pi_1(\tilde{X},\widetilde{x_0}))$ acts transitively on the fiber $(p^{-1}(x))$ for every $(x\in X)$. This implies that all sheets of the covering space $(\tilde{X})$ are equivalent under the action of the fundamental group, ensuring a uniform covering structure across (X).

ii. Galois Coverings (Han, 2010): Galois coverings extend the concept of regular coverings by ensuring that the action of the fundamental group $(\pi_1(X,x))$ on the fiber $\left(p^{-1}(x)\right)$ is both transitive and free. This additional condition implies that the covering $\left(p:\tilde{X}\rightarrow X\right)$ is a covering space that is also a Galois extension in the algebraic sense, leading to deeper connections between the topology of (X) and the algebraic structure of its covering spaces (Han, 2010).

7.4. APPLICATIONS TO ALGEBRAIC TOPOLOGY

Following are the applications of algebraic topology.

7.4.1. Van Kampen's Theorem

Van Kampen's theorem is a powerful tool in algebraic topology that allows the computation of the fundamental group of a space (X) by combining information from the fundamental groups of its open covering sets. Formally, let (X) be a topological space and $(\{U_\alpha\}_{\alpha\in A})$ be an open cover of (X) such that each intersection $(U_{\alpha_1} \cap U_{\alpha_2} \cap \ldots \cap U_{\alpha_n})$ is path-connected. Van Kampen's theorem states that if $(\pi_1(U_\alpha))$ denotes the fundamental group of (U_α) for each (α), then the fundamental group $(\pi_1(X))$ of (X) is the free product of the fundamental groups $(\pi_1(U_\alpha))$ amalgamated over their intersections (Zastrow, 1994).

This theorem is essential in handling spaces that can be decomposed into simpler, path-connected pieces, allowing for a systematic approach to calculating their fundamental groups. It finds applications in various areas of mathematics and physics, particularly in the study of spaces with complex structures that can be understood through their simpler constituent parts (Brown & Loday, 1987).

7.4.2. Higher Homotopy Groups

Higher homotopy groups extend the concept of the fundamental group to higher dimensions, capturing more intricate aspects of the topological structure of a space (X). For an integer $(n \geq 2)$, the (n)-th homotopy group $(\pi_n(X))$ is defined as the set of homotopy classes of continuous maps $(f: S^n \to X)$, where (S^n) denotes the dimensional sphere (Blanc, 1995). The group operation is given by the concatenation of maps and the homotopy equivalence relation (Papadima & Suciu, 2002).

These higher homotopy groups provide deeper insights into the connectivity and global structure of (X) beyond its fundamental group. They are fundamental invariants in algebraic topology, often used to distinguish between spaces that have the same fundamental group but differ in higher homotopy groups. Their computation and study shed light on the complexity of topological spaces and their behavior under continuous deformations (Dimca & Papadima, 2003).

7.4.3. Homotopy and Homology

Homotopy theory and homology theory are closely related fields in algebraic topology (Moore, 1954). While homotopy theory focuses on the deformation and equivalence of continuous mappings, homology theory studies algebraic structures derived from topological spaces, such as simplicial complexes or manifolds (Eilenberg & MacLane, 1945). Homotopy groups and homology groups provide complementary information about the space's structure: homotopy groups capture global connectivity and obstruction to continuous deformations, whereas homology groups quantify the cycles and boundaries in the space (Blakers, 1948), offering insights into its higher-dimensional structure and invariants. Together, they form powerful tools for analyzing and classifying topological spaces (Curtis, 1965).

ACTIVITY 7.1.

Explore covering spaces by investigating how they map onto a base space and how properties such as lifting and classification help in understanding the topological structure and symmetries of spaces.

SUMMARY

Chapter 7 delves into the intricate concepts of homotopy and the fundamental group, pivotal elements in the study of algebraic topology. The chapter begins with an exploration of the homotopy of paths, defining homotopy as a continuous deformation between paths within a topological space. This section provides concrete examples of homotopic and non-homotopic paths, highlighting their distinctions and practical implications. It further introduces homotopy equivalence and deformation, laying the groundwork for understanding homotopy in broader topological contexts and emphasizing the principle of homotopy invariance.

The second major section focuses on the fundamental group, a fundamental invariant in topology. It outlines the process of determining the fundamental group of simple spaces and presents methods for its calculation. Key properties of the fundamental group are discussed, including its algebraic structure and its role as a functor, which preserves the structure of topological spaces under continuous mappings. The chapter also examines the interplay between the fundamental group and covering spaces, elucidating how covering spaces can be used to compute and understand fundamental groups more comprehensively.

In the third section, covering spaces are defined and their basic properties are examined. The chapter elaborates on the lifting properties of covering spaces, essential for understanding how topological features can be projected and analyzed in different dimensions. The classification of covering spaces is also discussed, providing a framework for identifying and categorizing various types of coverings.

The final section applies the discussed concepts to algebraic topology. Van Kampen's Theorem is introduced as a powerful tool for computing the fundamental group of a space formed by the union of two spaces. The chapter then touches upon higher homotopy groups, extending the concept of the fundamental group to higher dimensions. Finally, the relationship between homotopy and homology is explored, connecting these two central ideas in algebraic topology.

Through a detailed examination of homotopy, the fundamental group, and covering spaces, this chapter equips readers with the necessary tools and understanding to tackle complex problems in algebraic topology. The discussions are augmented with rigorous mathematical definitions, theorems, and proofs, ensuring a comprehensive and technical understanding of these foundational concepts.

MULTIPLE CHOICE QUESTIONS

1. **The fundamental group of a topological space is:**
   a. A group that represents the number of holes in the space
   b. A group that classifies the space up to homotopy equivalence
   c. A group that measures the curvature of the space

d. A group that determines the volume of the space

2. **Which property is preserved under homotopy invariance?**
   a. Distance
   b. Area
   c. Continuous deformation
   d. Volume
3. **Covering spaces are used in topology primarily to:**
   a. Measure angles in a space
   b. Compute fundamental groups
   c. Calculate the volume of a space
   d. Define homotopy equivalence
4. **Van Kampen's Theorem is useful for:**
   a. Finding the diameter of a space
   b. Computing the fundamental group of a union of spaces
   c. Measuring the curvature of a surface
   d. Defining homotopy in higher dimensions
5. **Higher homotopy groups extend the concept of:**
   a. Fundamental group to higher dimensions
   b. Covering spaces to more complex spaces
   c. Homotopy equivalence to lower dimensions
   d. Volume calculation in topological spaces

REVIEW QUESTIONS

1. What is the definition of homotopy in the context of topological spaces?
2. Explain the concept of homotopy equivalence and how it relates to deformation.
3. Describe the fundamental group and its significance in topology.
4. How can covering spaces be used to compute fundamental groups?
5. Discuss the implications of Van Kampen's Theorem in algebraic topology.

Answers to Multiple Questions

1. (b) 2. (c) 3. (b) 4. (b) 5. (a)

REFERENCES

1. Adhikari, M. R. (2023). Homotopy Theory: Fundamental Group and Higher Homotopy Groups. In Basic Topology 3: Algebraic Topology and Topology of Fiber Bundles. Singapore: Springer Nature Singapore. (Vol. 2, pp. 27–161).
2. Ahiezer, D. N. (1974). Compact complex homogeneous spaces with solvable fundamental group. Mathematics of the USSR-Izvestiya, 8(1), 61.
3. Biagi, S., & Bonfiglioli, A. (2017). The existence of a global fundamental solution for homogeneous Hörmander operators via a global lifting method. Proceedings of the London Mathematical Society, 114(5), 855–889.
4. Blakers, A. L. (1948). Some relations between homology and homotopy groups. Annals of Mathematics, 49(2), 428–461.
5. Blanc, D. (1995). Higher homotopy operations and the realizability of homotopy groups. Proceedings of the London Mathematical Society, 3(1), 214–240.
6. Boggess, A., & Raich, A. (2009). A simplified calculation for the fundamental solution to the heat equation on the Heisenberg group. Proceedings of the American Mathematical Society, 137(3), 937–944.
7. Brown, R., & Loday, J. L. (1987). Van Kampen theorems for diagrams of spaces. Topology, 26(3), 311–335.
8. Cedzich, C., Geib, T., Stahl, C., Velázquez, L., Werner, A. H., & Werner, R. F. (2018). Complete homotopy invariants for translation invariant symmetric quantum walks on a chain. Quantum, 2, 95.
9. Conner, P. E., & Raymond, F. (1977). Deforming homotopy equivalences to homeomorphisms in aspherical manifolds. Bulletin of the American Mathematical Society, 83(1), 36–85.
10. Curtis, E. B. (1965). Some relations between homotopy and homology. Annals of Mathematics, 82(3), 386–413.
11. Dimca, A., & Papadima, S. (2003). Hypersurface complements, Milnor fibers and higher homotopy groups of arrangments. Annals of mathematics, 158(2), 473–507.
12. dos Santos, J. P. (2015). The homotopy exact sequence for the fundamental group scheme and infinitesimal equivalence relations. Algebr. Geom, 2(5), 535–590.
13. Dyer, M. N. (1978). On the essential height of homotopy trees with finite fundamental group. Compositio Mathematica, 36(2), 209–224.
14. Eilenberg, S., & MacLane, S. (1945). Relations between homology and homotopy groups of spaces. Annals of mathematics, 46(3), 480–509.
15. Gorbatsevich, V. V. (1986). Compact homogeneous spaces with a semisimple fundamental group. II. Siberian Mathematical Journal, 27(5), 660–669.
16. Groemer, H. (1986). Some basic properties of packing and covering constants. Discrete & computational geometry, 1(2), 183–193.
17. Han, S. E. (2010). Ultra regular covering space and its automorphism group. International Journal of Applied Mathematics and Computer Science, 20(4), 699–710.

18. Hardeman, R. (2018). Computing A-Homotopy Groups of Graphs Using Coverings and Lifting Properties. Computing, 2018, 09–21.
19. Hauschild, V. (1993). Deformations and the rational homotopy of the monoid of fiber homotopy equivalences. Illinois Journal of Mathematics, 37(4), 537–560.
20. Krivoshei, I. V. (1968). Calculation of the fundamental group for molecules of polymers and catenanes. Journal of Structural Chemistry, 9(2), 222–225.
21. Kruggel, B. (1998). Kreck-Stolz invariants, normal invariants and the homotopy classification of generalised Wallach spaces. The Quarterly Journal of Mathematics, 49(4), 469–485.
22. Li, T. Y., & Yorke, J. A. (1980). A simple reliable numerical algorithm for following homotopy paths. Analysis and computation of fixed points, 5, 73–91.
23. Maleki, M., Tonekaboni, S. A. M., & Abbasbandy, S. (2014). A homotopy analysis solution to large deformation of beams under static arbitrary distributed load. Applied Mathematical Modelling, 38(1), 355–368.
24. Marsden, J., & Scheurle, J. (1987). The construction and smoothness of invariant manifolds by the deformation method. SIAM journal on mathematical analysis, 18(5), 1261–1274.
25. Mironescu, P., & Van Schaftingen, J. (2021). Lifting in compact covering spaces for fractional Sobolev mappings. Analysis & PDE, 14(6), 1851–1871.
26. Moore, J. C. (1954). On homotopy groups of spaces with a single non-vanishing homology group. Annals of Mathematics, 59(3), 549–557.
27. Nori, M. V. (1976). On the representations of the fundamental group. Compositio Mathematica, 33(1), 29–41.
28. Papadima, S., & Suciu, A. I. (2002). Higher homotopy groups of complements of complex hyperplane arrangements. Advances in Mathematics, 165(1), 71–100.
29. Patera, J., Winternitz, P., & Zassenhaus, H. (1975). Continuous subgroups of the fundamental groups of physics. I. General method and the Poincaré group. Journal of Mathematical Physics, 16(8), 1597–1614.
30. Porter, T. (1987). Extensions, crossed modules and internal categories in categories of groups with operations. Proceedings of the Edinburgh Mathematical Society, 30(3), 373–381.
31. Rieser, A. (2021). Čech closure spaces: A unified framework for discrete and continuous homotopy. Topology and its Applications, 296, 107613.
32. Steinberg, B. (2003). The lifting and classification problems for subspaces of covering spaces. Topology and its Applications, 133(1), 15–35.
33. Thomas, E. (1964). Homotopy classification of maps by cohomology homomorphisms. Transactions of the American Mathematical Society, 111(1), 138–151.
34. Vakil, R., & Wickelgren, K. (2011). Universal covering spaces and fundamental groups in algebraic geometry as schemes. Journal de théorie des nombres de Bordeaux, 23(2), 489–526.
35. Zastrow, A. (1994). The second van-Kampen theorem for topological spaces. Topology and its Applications, 59(3), 201–232.

CHAPTER 8

Advanced Topics in Topology

LEARNING OBJECTIVES

After studying this chapter, you will be able to:

1. Understand the concept of homotopy and its significance in topological spaces.
2. Describe the fundamental group and its role in classifying topological spaces.
3. Calculate fundamental groups of simple spaces using various methods.
4. Explain the properties and functoriality of the fundamental group.
5. Analyze the relationship between fundamental groups and covering spaces.
6. Apply algebraic topology concepts such as Van Kampen's Theorem and higher homotopy groups.

KEY TERMS FROM THIS CHAPTER

Manifolds
Simplicial Complexes
Topological Groups
Topological Quantum Field Theory
Differentiable Manifolds
CW Complexes
Lie Groups

UNIT INTRODUCTION

Advanced topology encompasses diverse areas of study, each contributing unique perspectives and tools to understand complex geometric and algebraic structures. This chapter covers several key topics: manifolds, simplicial complexes, CW complexes, topological groups, and further applications in modern research (Mukherjee, 2006).

Manifolds serve as fundamental objects in differential geometry and topology, providing a local Euclidean structure that can be smoothly patched together to form a global geometric object. Simplicial complexes offer a combinatorial approach to topology, representing spaces as collections of simple geometric shapes called simplices. CW complexes generalize simplicial complexes, allowing for more flexible constructions and facilitating deeper insights into topological and homological properties (Davis & Kirk, 2001).

Topological groups integrate algebraic and topological structures, offering a framework to study symmetries and transformations of spaces. These groups find applications in physics, particularly in the formulation of gauge theories and the study of fundamental particles. Lastly, the chapter explores further applications such as topological data analysis, which provides tools to extract meaningful information from complex data sets, and topological quantum field theory, which investigates the topological aspects of quantum systems (Haefliger & Narasimhan, 2012).

8.1. MANIFOLDS

Manifolds are fundamental objects in mathematics and physics that generalize the concept of smooth curves and surfaces to higher dimensions (Figure 8.1). A manifold is a topological space that locally resembles Euclidean space near each point (Dubrovin et al., 2012). Formally, a manifold (M) of dimension (n) is a Hausdorff, second-countable space where every point has a neighborhood homeomorphic to an open subset of (R^n).

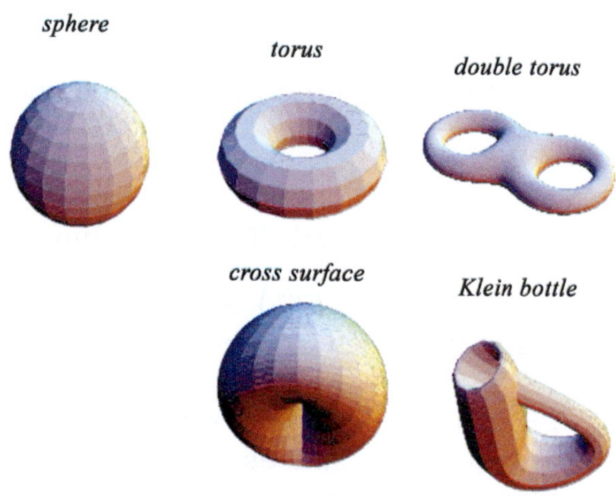

Figure 8.1. Non-intersecting closed surfaces in R^3 are examples of 2D manifolds such as a sphere, torus, double torus, cross surfaces and Klein bottle (Source: Wolfram, Creative Commons License)

8.1.1. Introduction to Manifolds

Intuitively, a manifold can be visualized as a space that looks flat (like Euclidean space) when viewed up close, but can be curved or otherwise structured when observed over larger scales (Trofimov, 2013). For instance, the surface of a sphere and a torus are examples of 2-dimensional manifolds embedded in 3-dimensional Euclidean space. These surfaces can be covered by coordinate patches that, when combined, provide a complete picture of the manifold's structure (Izenman, 2012).

8.1.2. Classification and Examples

Manifolds are classified based on various properties such as dimensionality, orientability, and the smoothness of their structure. They can be broadly categorized into (Chung et al., 2018):

i. Dimensionality: Manifolds can have any integer dimension (n), from zero-dimensional points to high-dimensional spaces relevant in physics and geometry.

ii. Orientability: Some manifolds can be oriented (e.g., they have consistent notions of clockwise and counterclockwise), while others are non-orientable (e.g., the Möbius strip).

iii. Smoothness: Differentiable manifolds are equipped with a smooth structure that allows for the definition of tangent spaces and differential forms, crucial for applications in differential geometry and physics (Lyakhovich et al., 2010).

Examples of manifolds include:

Remember

Remember that manifolds are spaces that locally resemble Euclidean space, crucial in fields like physics and geometry for modeling spaces with smooth structures.

i. Euclidean Spaces: (R^n) itself is a manifold, where each point has a neighborhood homeomorphic to an open subset of (R^n).

ii. Spheres and Tori: The $(n)-$dimensional sphere (S^n) and the $(n)-$torus (T^n) are fundamental examples of compact manifolds with and without boundary, respectively.

iii. Projective Spaces: Projective spaces (RP^n) and (CP^n) are quotient spaces that generalize Euclidean spaces by identifying points in a prescribed manner (Kreck & Schafer, 1984).

8.1.3. Differentiable Manifolds

Following is the explanation of differentiable manifolds.

8.1.3.1. Smooth Structures and Tangent Spaces

Differentiable manifolds extend the concept of manifolds by introducing a smooth structure that allows for the definition of tangent spaces and differential calculus. A smooth manifold (M) is equipped with an atlas of charts, each providing a smooth coordinate map to (R^n). These charts overlap smoothly on their intersections, ensuring a consistent transition between coordinates. Tangent spaces at each point of a smooth manifold are defined as the space of all derivations or directional derivatives at that point. They form a vector space that captures the local behavior of smooth curves passing through the point (Tyagi et l., 2013) (Figure 8.2).

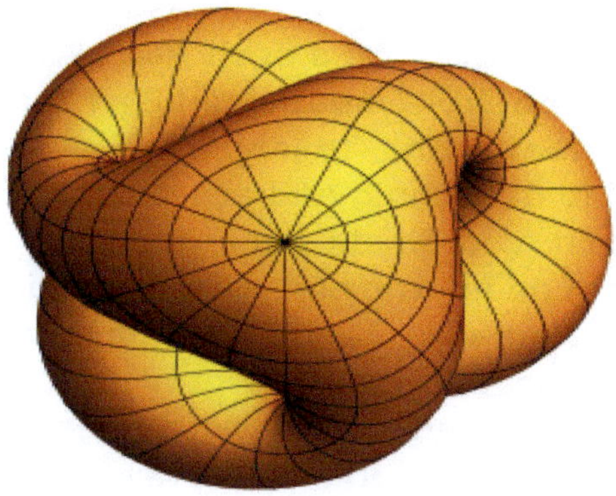

Figure 8.2. An immersion of the projective plane into R3 (Source: Emily Dumas, Creative Commons License)

8.1.3.2. Differential Forms and Integration

Differential forms on a differentiable manifold generalize the concept of differential forms from calculus. They are expressed as smooth sections of exterior powers of the cotangent bundle, providing a powerful framework for integration and flux calculations (Potepun, 2006). Integration of differential forms over manifolds yields global invariants that are insensitive to local coordinate choices, making them essential tools in geometric analysis and physics. Differentiable manifolds thus form a rich structure that merges local smoothness with global topological and differential properties, offering a comprehensive setting for studying geometric and analytic phenomena in higher dimensions (Gallier et al., 2020).

Did you know?

Did you know that manifolds are used extensively in physics to describe spacetime and in geometry to study surfaces and higher-dimensional spaces?

8.1.4. Applications in Physics and Geometry

Following are the applications in physics and geometry.

8.1.4.1. Manifolds in General Relativity

Manifold theory plays a pivotal role in Einstein's theory of General Relativity, where spacetime is modeled as a four-dimensional manifold equipped with a Lorentzian metric. This manifold represents the continuum of spacetime, where each point corresponds to an event and local physics is governed by the curvature of spacetime induced by matter and energy. The theory predicts gravitational effects as manifestations of the curvature of this manifold, emphasizing the interplay between geometry and physical phenomena (York, 1973).

8.1.4.2. Applications to Differential Geometry

In differential geometry, manifolds serve as the foundational setting for studying geometric structures such as curvature, connections, and geodesics. Differentiable manifolds allow for the development of tools like differential forms, tensors, and differential operators, which are essential in describing and analyzing geometric properties independent of specific coordinate systems (Joyce, 2014). These tools facilitate the study of geometric objects in intrinsic terms, leading to insights and results that are robust under smooth deformations and transformations. Moreover, manifolds provide a natural framework for understanding and formulating physical theories beyond General Relativity, including gauge theories in particle physics and geometric optics, highlighting their versatility and importance across diverse fields of mathematics and physics (Berger & Gostiaux, 2012).

Tip:

Simplicial complexes provide a combinatorial way to represent spaces, useful in computational geometry for tasks like mesh generation and topological data analysis.

8.2. SIMPLICIAL COMPLEXES

A simplicial complex is a fundamental object in algebraic topology that generalizes graphs and provides a combinatorial approach to studying topological spaces. Formally, a simplicial complex \(K \) is a collection of simplices that satisfies two properties (Baccini et al., 2022):

i. Closure under Faces: If (σ) is a simplex in (K), then every face of (σ) (including (σ) itself) must also be in (K).

ii. Intersection Property: The intersection of any two simplices in (K) is either empty or a face of each (Barbarossa & Sardellitti, 2020) (Figure 8.3).

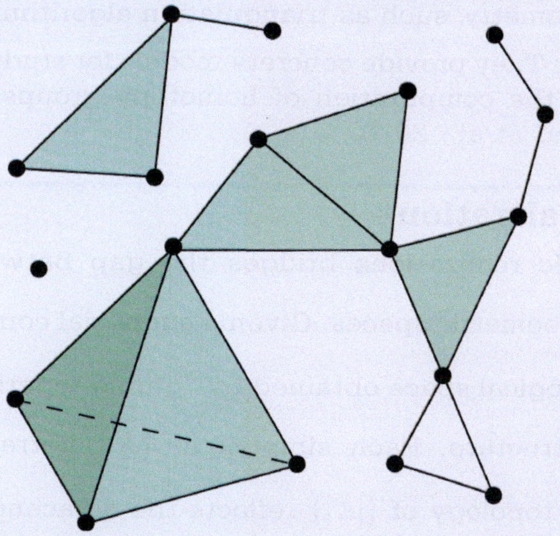

Figure 8.3. A simplicial 3-complex (Source: Trevor Good Child, Creative Commons License)

A simplex is a generalization of a triangle to higher dimensions. For instance, a 0-simplex is a vertex, a 1-simplex is an edge, a 2-simplex is a triangle, and so on. Examples of simplicial complexes include:

i. Graphs: A graph can be viewed as a 1-dimensional simplicial complex where vertices are 0-simplices and edges are 1-simplices (Kahle, 2014).

ii. Triangulations: Any polygonal region in the plane can be triangulated into 2-dimensional simplices, forming a simplicial complex.

iii. Higher-Dimensional Complexes: These include higher-dimensional simplicial complexes used to model more complex spaces, such as surfaces and higher-dimensional manifolds (Jonsson, 2003).

Simplicial Homology: Simplicial homology is a fundamental tool in algebraic topology that associates algebraic objects (homology groups) to simplicial complexes. It captures essential topological features such as holes, tunnels, and voids in spaces represented by simplicial complexes. The (n)-th simplicial homology group $(H_n(K))$ of a simplicial complex (K) measures $(n)-$dimensional "cycles" (chains that do not bound any higher-dimensional chain) modulo "boundaries" (chains that are boundaries of higher-dimensional chains). This construction provides a combinatorial method to compute and understand topological invariants of spaces represented by simplicial complexes (Jiang & Omer, 2007).

Simplicial complexes are crucial in various applications, including:

i. Topological Data Analysis: They are used to analyze and visualize data by constructing simplicial complexes from point cloud data and computing persistent homology.

ii. Computational Geometry: Simplicial complexes form the basis for algorithms in computational geometry, such as triangulation algorithms and mesh generation.

iii. Homotopy Theory: They provide concrete models for studying spaces in homotopy theory, aiding in the computation of homotopy groups and other homotopical invariants (Serrano et al., 2020).

8.2.1. Geometric Realizations

The concept of geometric realizations bridges the gap between abstract simplicial complexes and concrete geometric spaces. Given a simplicial complex (K), its geometric realization $(|K|)$ is a topological space obtained by "gluing together" simplices according to their combinatorial structure. Each simplex in (K) corresponds to a geometric simplex in $(|K|)$, and the topology of $(|K|)$ reflects the adjacency relations specified by the simplicial complex.

Triangulations are a specific type of geometric realization where a topological space is decomposed into simplices. For example, a polygon can be triangulated into triangles, a polyhedron into tetrahedra, and higher-dimensional spaces into higher-dimensional simplices. Triangulations not only facilitate the visualization and computation of geometric objects but also provide a combinatorial understanding of their topological properties.

Combinatorial structures underlying geometric realizations involve the study of how simplicial complexes encode the topological features of spaces. This includes understanding how different arrangements of simplices (faces and edges) give rise to various topological properties such as connectivity, compactness, and homotopy equivalence. Geometric realizations play a crucial role in applications such as:

1. Computational Geometry: Algorithms for mesh generation, surface reconstruction, and finite element analysis often rely on triangulations and their geometric realizations.
2. Topological Data Analysis: Persistent homology, a powerful tool for analyzing data sets, utilizes geometric realizations of simplicial complexes to extract and quantify topological features.

In essence, geometric realizations of simplicial complexes provide a versatile framework for both theoretical investigations and practical applications in mathematics, computer science, and engineering. They unify combinatorial and geometric approaches to understanding complex structures and their inherent topological properties (Biasotti et al., 2008).

8.2.2. Applications in Computational Geometry

Simplicial complexes and their geometric realizations are foundational in computational geometry for solving a variety of problems. Algorithms based on simplicial complexes are used for tasks such as mesh generation, surface reconstruction, and spatial partitioning. Triangulations, in particular, provide a robust method for discretizing geometric objects into simplices, which can then be efficiently processed and analyzed computationally (Dong et al., 2022).

In data science and analysis, simplicial complexes play a crucial role in representing and analyzing complex data sets. Topological data analysis (TDA) utilizes simplicial complexes to capture and quantify the topological features of data, revealing patterns and structures that are not apparent in traditional Euclidean or metric spaces. Persistent homology, a key technique in TDA, computes topological invariants from simplicial complexes to identify significant features and extract meaningful insights from noisy or high-dimensional data.

These applications underscore the versatility and power of simplicial complexes in computational geometry and data analysis. By combining geometric and combinatorial approaches, simplicial complexes provide a unified framework for tackling complex problems in mathematics, computer science, and various scientific disciplines. Their ability to bridge theoretical concepts with practical algorithms makes them indispensable tools for understanding and manipulating geometric and topological structures in modern computational environments (Musin, 2006).

8.3. CW COMPLEXES

CW complexes are a class of topological spaces that generalize simplicial complexes, providing a flexible framework for studying spaces with more intricate topological properties. They are constructed by attaching cells of varying dimensions to a base space in a systematic manner. The term "CW" stands for "closure-finite" and "weak topology," highlighting two key aspects of their construction:

Cells and Attachments: A CW complex (X) is built iteratively from lower-dimensional cells called (n)-cells. Starting with $(0)-$cells (points), one attaches $(1)-$cells (edges) to form 1-dimensional structures, then $(2)-$cells (disks) to form 2-dimensional surfaces, and so forth. This process continues indefinitely, constructing higher-dimensional cells by attaching them along the boundaries of previously attached cells (Figure 8.4).

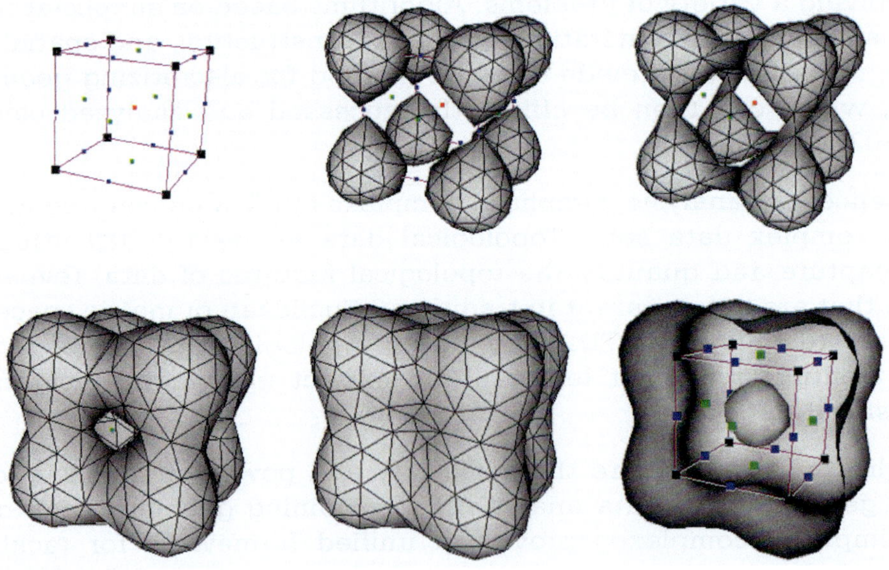

Figure 8.4. The CW-complex of a torus (Source: John C. Hart, Creative Commons License)

Properties and Characteristics: The construction of a CW complex (X) ensures that it inherits both the local properties of cells and the global topological properties of the attaching maps. This allows for a rich interplay between combinatorial structure and topological properties such as connectivity, compactness, and homotopy type. Unlike simplicial complexes, which are rigidly defined by their combinatorial structure, CW complexes are more flexible and can model a wider range of topological spaces, including those with non-trivial homotopy groups and more intricate geometrical features (Hart, 2005).

CW complexes find applications across various fields, including:

1. Algebraic Topology: They serve as a primary tool for computing homology and cohomology groups, providing insights into the underlying topological structure of spaces.
2. Differential Topology: In differential geometry, CW complexes are used to study smooth and differentiable manifolds, providing a geometric framework for understanding curvature and differential forms.
3. Physics and Geometry: In theoretical physics, CW complexes are employed in string theory and quantum field theory to model space-time and gauge theories, facilitating the study of complex physical phenomena through geometric abstraction.

In conclusion, CW complexes offer a versatile and powerful approach to studying topological spaces, combining geometric intuition with rigorous mathematical structure. Their construction allows for a nuanced exploration of spaces with varying degrees of complexity, making them indispensable tools in modern mathematics and its applications to other scientific disciplines (Mimura et al., 1971).

8.3.1. Cellular Homology

Homology Theories in CW Complexes: Cellular homology is a fundamental aspect of algebraic topology that studies the topological invariants of CW complexes. It provides a method for computing homology groups, which characterize the "holes" or higher-dimensional voids within a space. In a CW complex (X), cellular homology assigns an abelian group $(H_n(X))$ to each dimension (n), capturing cycles (closed chains) and boundaries (chains that are boundaries of higher-dimensional chains).

Relationship to Simplicial Homology: While both cellular and simplicial homology are methods for computing homology groups, they differ in their construction and applicability. Simplicial homology relies on triangulations of spaces into simplices, whereas cellular homology utilizes the cell structure inherent in CW complexes. The boundary operators in cellular homology are determined by the attaching maps of cells, which simplifies computations and often provides more intuitive insights into the topological features of a space.

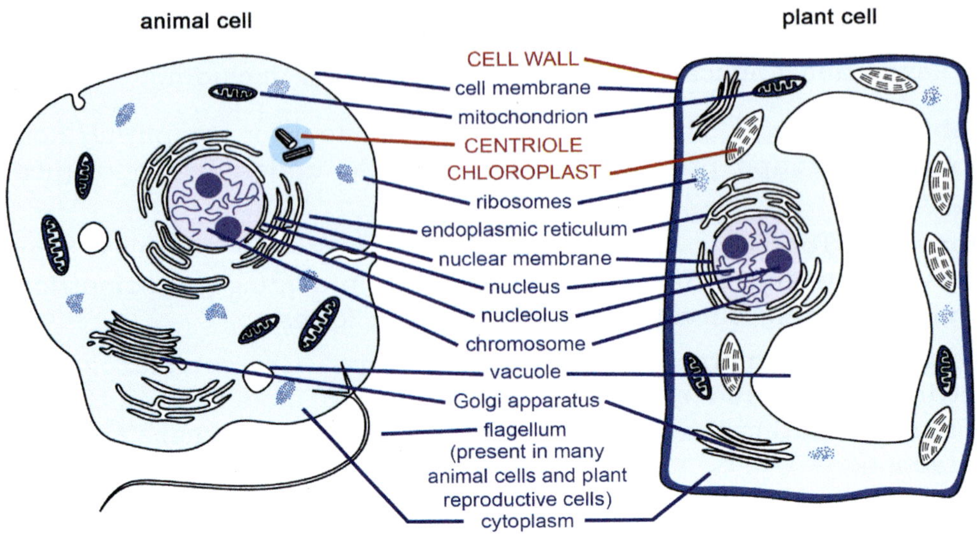

Figure 8.5. At the cellular and molecular levels, all living things are fundamentally alike. These basic similarities are most easily explained by evolution in Homology (Source: Simakov, Creative Commons License)

Applications and Comparisons: Cellular homology is particularly advantageous in handling spaces that are naturally modeled by CW complexes, such as manifolds and certain types of topological spaces with specific cellular decompositions. It complements simplicial homology by offering alternative approaches to computing homology groups, often revealing different aspects of the underlying space. Moreover, cellular homology theories extend to more advanced topics in algebraic topology, including spectral sequences and the computation of persistent homology in topological data analysis (Alayrangues et al., 2015).

8.3.2. Examples and Applications

Classification of Surfaces: One of the prominent applications of CW complexes is in the classification of surfaces. Surfaces can be categorized based on their genus (number of handles) and orientability. For instance, the sphere (S^2), the torus (T^2), and the projective plane (RP^2) can be represented and studied using CW complexes. The cellular structure of these surfaces provides a systematic way to understand their homology and homotopy properties, facilitating their classification and deeper geometric analysis.

Homotopy Theory and CW Approximations: CW complexes play a crucial role in homotopy theory by providing approximations of more general spaces. Through CW approximations, any topological space can be "approximated" by a CW complex, capturing its essential homotopy features. This method allows for simplifying the study of homotopy equivalences and fundamental groups of spaces, as well as computing higher homotopy groups. CW complexes serve as foundational tools in homotopy theory, offering insights into the behavior of continuous mappings and the classification of maps up to homotopy equivalence (Shankar, 1977).

Applications in Mathematics and Beyond: Beyond pure mathematics, CW complexes find applications in theoretical physics, computer science, and data analysis. They provide a versatile framework for modeling and analyzing complex systems and structures, leveraging their combinatorial and geometric properties to solve problems in diverse fields. Their utility extends to fields such as robotics, where they are used to model configurations and movement spaces, and in algorithms for topological data analysis, which extracts meaningful information from complex data sets using homological methods.

In essence, CW complexes are indispensable tools in mathematics and its applications, offering a unified approach to understanding and exploring the topology of spaces through their cell structure and homotopy properties. Their applications continue to evolve, driving advancements in various scientific disciplines and theoretical frameworks (Dow, 1988).

8.4. TOPOLOGICAL GROUPS

A topological group is a mathematical structure that combines the properties of a group and a topological space in a compatible manner. Formally, it is a group (G) equipped with a topology such that both the group operations (multiplication and inversion) and the topological operations (open sets, continuity) interact smoothly. Specifically, the group multiplication $(G \times G \to G)$ and the inversion $(G \to G)$ are required to be continuous maps with respect to the topology on (G).

The fundamental properties of topological groups include:

1. Compatibility of Group Operations: The group multiplication $((g,h) \mapsto gh)$ and inversion $(g \mapsto g^{-1})$ are continuous maps on (G).
2. Identity and Inverses: The identity element $(e \in G)$ has an open neighborhood in (G), and every element $(g \in G)$ has an open neighborhood (U) such that $(gU \subseteq U)$ and $(Ug^{-1} \subseteq U)$.
3. Topological Group Structure: The topology on (G) is typically required to be Hausdorff to ensure separation of points (Uspenski, 1990).

8.4.1. Group Actions and Applications

In the context of topological groups, group actions on topological spaces play a significant role. A group action $(G \times X \to X)$ of a topological group (G) on a topological space (X) is a continuous map that respects the group structure. This notion extends symmetry and transformation concepts from algebraic groups to topological settings, allowing for the study of symmetries and invariant properties of spaces.

Topological groups find applications across various branches of mathematics and beyond:

i. Representation Theory: They provide a framework for studying representations of groups on vector spaces, yielding insights into algebraic structures and symmetry.

ii. Dynamics and Ergodic Theory: Group actions on dynamical systems lead to the study of invariant measures and ergodic properties, crucial in understanding chaotic behavior and stability.

iii. Number Theory and Geometry: In arithmetic geometry, topological groups such as the adele ring and p-adic groups play a central role in understanding number fields and their geometrical properties (Bekka & Mayer, 2000).

8.4.2. Lie Groups and Geometry

Lie groups are a special class of topological groups that are also smooth manifolds. They combine the algebraic structure of groups with the differential structure of manifolds, making them essential in differential geometry and theoretical physics. Lie groups provide a framework for studying symmetries and transformations that preserve geometric structures, such as curvature and metric properties.

The study of Lie groups and their actions on manifolds underpins:

i. Geometric Structures: Lie groups act as transformation groups on geometric structures, leading to the study of homogeneous spaces and symmetric spaces.

ii. Physics: In theoretical physics, Lie groups are used to formulate gauge theories, describe particle symmetries, and model spacetime symmetries in general relativity (Adhikari & Adhikari, 2022).

8.5. FURTHER APPLICATIONS AND RESEARCH DIRECTIONS

Following are the applications and research directions.

8.5.1. Topological Data Analysis

Topological data analysis is an emerging field that applies concepts from algebraic topology to analyze complex data sets. By converting data points into a topological space and studying its shape and connectivity, TDA provides insights into the underlying structure and patterns that traditional methods might overlook. Techniques such as persistent homology capture robust features of data across multiple scales, revealing essential topological characteristics that can guide decision-making in fields ranging from biology and neuroscience to computer vision and materials science. TDA's ability to handle noise and uncertainty in data makes it a powerful tool for extracting meaningful information and understanding the intrinsic geometry of high-dimensional data sets (Sizemore et al., 2019).

8.5.2. Topological Quantum Field Theory

TQFT is a branch of quantum field theory that focuses on topological aspects rather than metric properties. It provides a mathematical framework for studying quantum field theories that are invariant under differentiable and topological transformations. TQFT has applications in theoretical physics, particularly in understanding topological phases of matter, quantum computing, and the classification of quantum states based on topological invariants. By abstracting away metric details and focusing on topological invariants, TQFT provides insights into universal properties of physical systems and has implications for quantum gravity and the study of exotic phases of matter (Alexandrov et al., 1997).

8.5.3. Open Problems and Current Research

In topology, several intriguing open problems and current research directions include:

1. Homotopy Type Theory: Exploring connections between homotopy theory and constructive mathematics, especially in the context of type theory.

2. Higher Homotopy Theory: Investigating higher-dimensional analogs of fundamental group theory, such as higher homotopy groups and their applications.
3. Topological Robotics: Applying topological methods to the study of robot motion planning and configuration spaces.
4. Topological Phases of Matter: Understanding novel phases of matter that emerge from topological considerations in condensed matter physics (Jongeneel & Moulay, 2023).

8.5.4. Current Research

Recent advancements focus on developing computational tools for TDA, refining methods for computing persistent homology, and applying TQFT to model and predict experimental outcomes in quantum materials (Vanhoucke, 2010). Furthermore, interdisciplinary collaborations between mathematicians, physicists, computer scientists, and biologists continue to expand the frontiers of topology's applications, integrating theoretical insights with practical challenges in data analysis, quantum information, and complex systems (Neumann, 1994).

In conclusion, topology's impact extends beyond traditional mathematical domains, influencing diverse fields through its rigorous methods and deep connections to geometry, algebra, and physics. As research continues to uncover new connections and applications, topology remains at the forefront of innovation, driving discoveries and advancing our understanding of complex systems and fundamental principles in science and technology (Egenhofer & Sharma, 1993).

ACTIVITY 8.1.

Explore CW complexes by constructing examples and understanding how their cellular structure aids in computing homology, which measures holes and voids in spaces.

SUMMARY

Chapter 8 delves into some of the most intricate and fascinating areas in topology, presenting an in-depth analysis of advanced topics that play a crucial role in both theoretical and applied mathematics. The chapter begins with an introduction to manifolds, covering their fundamental definitions and classifications. It explores differentiable manifolds, emphasizing their importance and applications in physics and geometry. The discussion extends to various examples and the specific characteristics that distinguish different types of manifolds.

The chapter then transitions to simplicial complexes, examining their geometric realizations and applications in computational geometry. This section highlights how simplicial complexes serve as powerful tools in various computational problems, providing a bridge between discrete and continuous mathematics. The analysis includes a detailed study of their structure and the methods used to construct and manipulate them.

Next, the chapter explores CW complexes, focusing on cellular homology and its applications. CW complexes are pivotal in algebraic topology, and their examination provides insights into the methods used to compute homological invariants. Examples and applications illustrate the practical relevance of these complexes in solving topological problems.

The discussion on topological groups introduces the concept of group actions and their applications, followed by an exploration of Lie groups and their geometric properties. Topological groups are essential in understanding symmetry and continuous transformations in topology, and this section elucidates their theoretical and practical significance.

The chapter concludes with an exploration of further applications and research directions in topology. Topics such as topological data analysis and topological quantum field theory are introduced, highlighting their cutting-edge applications in various scientific fields. The final section addresses open problems and current research, providing a glimpse into the ongoing advancements and unresolved questions in topology.

Throughout the chapter, rigorous mathematical concepts are presented alongside practical examples and applications, ensuring a comprehensive understanding of these advanced topics. Equations and formal definitions are included to support the theoretical discussions, making this chapter a vital resource for students and researchers seeking to deepen their knowledge of advanced topological methods and their applications.

MULTIPLE CHOICE QUESTIONS

1. **What is a differentiable manifold?**
   a. A space that is homeomorphic to a Euclidean space.
   b. A manifold with a differentiable structure.

c. A manifold that cannot be embedded in a higher-dimensional space.
d. A manifold that is compact and connected.

2. Which of the following is true about simplicial complexes?

a. They are always one-dimensional.
b. They have applications in computational geometry.
c. They cannot be used to study homology.
d. They are not used in topology.

3. CW complexes are primarily used to:

a. Study differential equations.
b. Compute homological invariants.
c. Analyze numerical algorithms.
d. Perform statistical analysis.

4. Lie groups are important in topology because they:

a. Provide examples of non-differentiable manifolds.
b. Represent discrete symmetries in geometric structures.
c. Describe continuous symmetries and transformations.
d. Are used exclusively in algebraic geometry.

5. Topological data analysis is a field that:

a. Focuses on solving differential equations.
b. Studies the topological properties of quantum fields.
c. Analyzes the shape of data using topological methods.
d. Examines the algebraic structures of data sets.

REVIEW QUESTIONS

1. Explain the fundamental concept of a manifold and provide examples of different types of manifolds.
2. Describe the process of constructing a simplicial complex and its applications in computational geometry.
3. Discuss the significance of cellular homology in CW complexes and provide examples of its applications.
4. Define topological groups and explain the importance of group actions in topology.
5. What are the current research directions in topological quantum field theory and topological data analysis?

Answers to Multiple Questions

1. (b) 2. (b) 3. (b) 4. (c) 5. (c)

REFERENCES

1. Adhikari, A., & Adhikari, M. R. (2022). Brief History of Topological Groups, Manifolds and Lie Groups. In *Basic Topology 2: Topological Groups, Topology of Manifolds and Lie Groups*. Singapore: Springer Nature Singapore. (Vol. 3, pp. 355–364).
2. Alayrangues, S., Damiand, G., Lienhardt, P., & Peltier, S. (2015). Homology of cellular structures allowing multi-incidence. *Discrete & Computational Geometry*, *54*, 42–77.
3. Alexandrov, M., Schwarz, A., Zaboronsky, O., & Kontsevich, M. (1997). The geometry of the master equation and topological quantum field theory. *International Journal of Modern Physics A*, *12*(7), 1405–1429.
4. Baccini, F., Geraci, F., & Bianconi, G. (2022). Weighted simplicial complexes and their representation power of higher-order network data and topology. *Physical Review E*, *106*(3), 34319.
5. Barbarossa, S., & Sardellitti, S. (2020). Topological signal processing over simplicial complexes. *IEEE Transactions on Signal Processing*, *68*, 2992–3007.
6. Bekka, M. B., & Mayer, M. (2000). *Ergodic theory and topological dynamics of group actions on homogeneous spaces*. Cambridge University Press. (Vol. 269, pp. 65–69).
7. Berger, M., & Gostiaux, B. (2012). *Differential Geometry: Manifolds, Curves, and Surfaces: Manifolds, Curves, and Surfaces.* Springer Science & Business Media. (Vol. 115, pp. 98–100).
8. Biasotti, S., De Floriani, L., Falcidieno, B., Frosini, P., Giorgi, D., Landi, C., ... & Spagnuolo, M. (2008). Describing shapes by geometrical-topological properties of real functions. *ACM Computing Surveys (CSUR)*, *40*(4), 1–87.
9. Chung, S., Lee, D. D., & Sompolinsky, H. (2018). Classification and geometry of general perceptual manifolds. *Physical Review X*, *8*(3), 31003.
10. Davis, J. F., & Kirk, P. (2001). *Lecture notes in algebraic topology.* Providence: American Mathematical Society (Vol. 35, pp. 367).
11. Dong, Z., Lin, H., & Chen, J. (2022). Computational Topology and its Applications in Geometric Design. *Recent Patents on Engineering*, *16*(5), 3–18.
12. Dow, A. (1988). An introduction to applications of elementary submodels to topology. In *Topology Proc* (Vol. 13, pp. 17–72).
13. Dubrovin, B. A., Fomenko, A. T., & Novikov, S. P. (2012). *Modern geometry—methods and applications: Part II: The geometry and topology of manifolds.* Springer Science & Business Media. (Vol. 104, pp. 65–70).
14. Egenhofer, M., & Sharma, J. (1993). Assessing the consistency of complete and incomplete topological information. *Geographical Systems*, *1*(1), 47–68.
15. Gallier, J., Quaintance, J., Gallier, J., & Quaintance, J. (2020). Integration on Manifolds. *Differential Geometry and Lie Groups: A Second Course*, 6, 211–263.
16. Haefliger, A., & Narasimhan, R. (Eds.). (2012). *Essays on topology and related topics: Memoires dedies a Georges de Rham*. Springer Science & Business Media. (Vol. 2, pp. 14–18).

17. Hart, J. C. (2005). Using the CW-complex to represent the topological structure of implicit surfaces and solids. In *ACM SIGGRAPH 2005 Courses* (Vol. 3, pp. 52–55).

18. Izenman, A. J. (2012). Introduction to manifold learning. *Wiley Interdisciplinary Reviews: Computational Statistics*, *4*(5), 439–446.

19. Jiang, B., & Omer, I. (2007). Spatial topology and its structural analysis based on the concept of simplicial complex. *Transactions in GIS*, *11*(6), 943–960.

20. Jongeneel, W., & Moulay, E. (2023). *Topological Obstructions to Stability and Stabilization: History, Recent Advances and Open Problems.* Springer Nature. (Vol. 2, pp. 132).

21. Jonsson, J. (2003). On the topology of simplicial complexes related to 3-connected and Hamiltonian graphs. *Journal of Combinatorial Theory, Series A*, *104*(1), 169–199.

22. Joyce, D. (2014). An introduction to d-manifolds and derived differential geometry. *Moduli spaces, LMS Lecture Notes*, *411*, 230–281.

23. Kahle, M. (2014). Topology of random simplicial complexes: a survey. *AMS Contemp. Math*, *620*, 201–222.

24. Kreck, M., & Schafer, J. A. (1984). Classification and stable classification of manifolds: some examples. *Commentarii Mathematici Helvetici*, *59*, 12–38.

25. Lyakhovich, S. L., Mosman, E. A., & Sharapov, A. A. (2010). Characteristic classes of Q-manifolds: classification and applications. *Journal of Geometry and Physics*, *60*(5), 729–759.

26. Mimura, M., Nishida, G., & Toda, H. (1971). Localization of CW-complexes and its applications. *Journal of the Mathematical Society of Japan*, *23*(4), 593–624.

27. Mukherjee, B. (2006). Advanced Topics in Virtual-Topology Optimization. *Optical WDM Networks*, 4, 433–477.

28. Musin, O. R. (2006). On some problems of computational geometry and topology. In *Global Analysis-Studies and Applications V.* Berlin, Heidelberg: Springer Berlin Heidelberg. (Vol. 2, pp. 57–80).

29. Neumann, J. (1994). The topological information content of a map/an attempt at a rehabilitation of information theory in cartography. *Cartographica: The International Journal for Geographic Information and Geovisualization*, *31*(1), 26–34.

30. Potepun, A. V. (2006). Integration of differential forms on manifolds with locally-finite variations. *Journal of Mathematical Sciences*, *139*, 6457–6478.

31. Serrano, D. H., Hernández-Serrano, J., & Gómez, D. S. (2020). Simplicial degree in complex networks. Applications of topological data analysis to network science. *Chaos, Solitons & Fractals*, *137*, 109839.

32. Shankar, R. (1977). Applications of topology to the study of ordered systems. *Journal de Physique*, *38*(11), 1405–1412.

33. Sizemore, A. E., Phillips-Cremins, J. E., Ghrist, R., & Bassett, D. S. (2019). The importance of the whole: topological data analysis for the network neuroscientist. *Network Neuroscience*, *3*(3), 656–673.

34. Trofimov, V. V. (2013). *Introduction to Geometry of Manifolds with Symmetry.* Springer Science & Business Media (Vol. 270, pp. 65–70).

35. Tyagi, H., Vural, E., & Frossard, P. (2013). Tangent space estimation for smooth embeddings of Riemannian manifolds®. *Information and Inference: A Journal of the IMA*, *2*(1), 69–114.

36. Uspenskiĭ, V. V. (1990). Topological groups and Dugundji compacta. *Mathematics of the USSR-Sbornik*, *67*(2), 555.

37. Vanhoucke, M. (2010). Using activity sensitivity and network topology information to monitor project time performance. *Omega*, *38*(5), 359–370.

38. York Jr, J. W. (1973). Conformally invariant orthogonal decomposition of symmetric tensors on Riemannian manifolds and the initial-value problem of general relativity. *Journal of Mathematical Physics*, *14*(4), 456–464.

Index

V

W

Printed and bound by CPI Group (UK) Ltd, Croydon, CR0 4YY
18/09/2026
14965063-0001